Programming as if People Mattered

Programming as if People Mattered

Friendly Programs,

Software Engineering,

and Other Noble Delusions

Nathaniel S. Borenstein

Princeton University Press

Princeton, New Jersey

Copyright © 1991 by Princeton University Press

Published by Princeton University Press,
41 William Street, Princeton, New Jersey 08540

In the United Kingdom: Princeton University Press,
Chichester, West Sussex

Library of Congress Cataloging-in-Publication Data

Borenstein, Nathaniel S.
Programming as if people mattered: friendly programs, software engineering, and
other noble delusions / Nathaniel S. Borenstein.
      p.   cm.
      Includes bibliographical references.
      ISBN 0-691-08752-0 (CL)
      ISBN 0-691-03763-9 (PB)
      1. Software engineering. 2. User interfaces (Computer systems)
I. Title.
QA76.758.B67 1991
004'.01'9–dc20   91-199969 CIP

Third printing, for the Princeton paperback edition, 1994

Princeton University Press books are printed on acid-free paper and meet the
guidelines for permanence and durability of the Committee on Production Guidelines
for Book Longevity of the Council on Library Resources

Printed in the United States of America

10 9 8 7 6 5 4 3

To

Perry Smith,
Harold Kasimow,
Henry Walker,
Jim Morris,
and Ezra Sensibar

for doing their best to teach me to
think critically,
act ethically,
and live with integrity

Contents

Part Four

The Golden Path: The Road to Human-Oriented
Software Engineering   137

Preface

To many highly trained computer specialists, writing computer programs is easy, as long as they don't have to get involved in questions of *user interface*—the actual details of the mechanisms by which human beings interact with computers. User-interface programming, it is widely perceived, is a quagmire in which expert programmers can become hopelessly and unproductively bogged down almost indefinitely due to the fundamental perversity, or at least unpredictability, of human nature.

To many people, including psychologists, industrial designers, and others, it is often painfully obvious how to improve the user interface of most programs. However, they generally tend to avoid actually fixing anything because it isn't their area of training. Those who do try their hand at it often find that the technical details of programming can be a lot harder than they look.

The topic of this book is the fundamental difficulty of writing high-quality computer programs that are easy for users to work with and understand. The main problem is that two rather divergent communities have evolved to address two different facets of this problem.

On the one hand, specialists in user-centered software design, including psychologists and others, have accumulated a body of knowledge about how to build usable programs. On the other hand, software engineers have made progress in understanding how to structure the software development process to increase the likelihood that it will result in reliable, robust, maintainable code.

This book was motivated by my personal observation that the solutions these two communities have found appear to be inadequate and, in some degree, mutually incompatible. Software engineering methods, even when they are successful, are not distinguished for producing high-quality user interfaces, and user-interface specialists are not well known for producing reliable and easily maintained programs. The partial and incomplete solutions that have been produced by these two diverging cultures are rarely, if ever, brought together in a single software development project.

What you are holding in your hands, then, is an attempt to bridge a wide and regrettable gap. The key question is a simple one: is it possible to develop systems in such a way that the product is both well designed for the user and well engineered as computer software? It is my hope that this book will help point the way toward an affirmative answer to this question.

Arriving at that answer will not be easy, however. Between the software engineers and the human-interface specialists there is already a substantial gap of knowledge and even terminology. Relatively few people are familiar with both of these specializations, and accordingly a substantial portion of this book is a summary of the state of the art in the two disciplines.

In accordance with the interdisciplinary nature of the problem, this book has been written with a number of different audiences in mind:

> *Managers of user-interface programming projects* are possibly the most important audience for this book. Such projects are routinely ruined or made much less successful by the mistakes of managers who fail to understand either the requirements of end-users or the realities of software engineering. The combination of these two kinds of expertise is rare to begin with, and even rarer in managers, but no one person has more need for both specialties than the manager in charge of the project. This book is written, first and foremost, to help managers figure out how to balance the often conflicting requirements of software engineering and user-centered software design.
>
> *Students of programming* who are contemplating a career in user-interface software development are, in fact, an audience whose needs resemble those of managers. Those who hope to specialize in user interfaces will need to learn how to look at problems from both a user-centered and a software engineering perspective, and this book might serve as a starting point in that interdisciplinary education.

Software engineering researchers who are interested in extending their results and methodologies to the area of user interfaces may find this book somewhat provocative. A major premise of the book is that a user-centered approach renders invalid a substantial portion of the techniques that work well for other parts of software engineering. This book might thus be regarded by the software engineering community as a challenge to try to incorporate modern user-centered design practices more completely into engineering methodologies.

Finally, *human-computer interaction researchers* might find this book intriguing, if not exactly enlightening, as a kind of "report from the trenches." Most of the user-interface principles described here will not come as any great surprise to the user-interface specialist, but the realities and practices of software engineering may be less familiar. My recommendations for reconciling the two will doubtless be unpopular in many quarters.

While user-interface specialists would no doubt benefit substantially from a serious overview of software engineering principles and practices, the emphasis of this book is largely in the other direction. That is, software engineers will learn more from this book about user interfaces than user-interface designers will learn about software engineering. This emphasis on teaching software engineers about user interfaces simply reflects a desire to have a more concrete effect: it is the software engineers who, ultimately, build most computer programs, and therefore it is by broadening their horizons through a more interdisciplinary education that the greatest benefit is likely to be obtained. If user-interface specialists keep their heads in the clouds where certain practical matters are concerned, this will do less genuine harm than if those who actually write software remain ignorant of what has been learned about user-interface design.

It should also be noted that while the focus of this book is user-interface development, it inevitably branches somewhat farther afield. Sometimes, though not always, the best principles for user-interface development also apply to software development in general.

Acknowledgments

Many people—probably more than I have remembered to credit here—have contributed to this book in many ways.

I am indebted to all the people who read and commented on drafts of some or all of the manuscript, including Mike Bianchi, Trina Borenstein, Marc Donner, Bob Glickstein, Chris Haas, Jim Hollan, Brad Myers, David Reiner, Steve Webster, and some anonymous reviewers. Particularly helpful and supportive were Lilya Lorrin, Lauren Oppenheim, and the rest of the staff at Princeton University Press.

I am grateful to all of those who contributed anecdotes, horror stories, or other references, many of which I ended up using in this book. Among those in this category are Joe Bates, Ernst Biersack, Larry Campbell, Warren Carithers, Brian Coan, Scott Deerwester, Bob Glickstein, David Copp, Ralph Hill, Nat Howard, Ingemar Hulthage, Bonnie John, Craig Knoblock, Tony Kwong, Jill Larkin, Lynne Reder, Alex Rudnicky, Lee Sproull, Mary Shaw, Dave Sincoskie, and Jamie Zawinski.

Students in two classes that I taught were subjected to early drafts of this book. The first was a one-day tutorial at the 1988 CIPS conference in Edmonton, Alberta, and the second was a special topics course at Carnegie Mellon University in the spring of 1989. In both cases, the students provided helpful feedback that is, I hope, reflected in the final version of the book.

I am particularly grateful to Bell Communications Research, Carnegie Mellon University, and IBM. These three organizations provided me with two wonderful work environments in which I was first able to learn the lessons I have tried to convey in this book, and then able to find the time to write them down.

I wish to give special thanks to Al Buzzard, Bob Kraut, Jim

Morris, Eric Nussbaum, Stu Personick, Jonathan Rosenberg, Al Spector, and Steve Weinstein, for running the enlightened research organizations within which I have been privileged to work during the past few years.

Finally, I am, as always, indebted to my wife and children for keeping me sane, motivated, and happy throughout the writing of this book.

Programming as if People Mattered

Part One

The Journey to the East: Can Software Engineers Build User Interfaces?

It was my destiny to join in a great experience. Having had the good fortune to belong to the League, I was permitted to be a participant in a unique journey. What wonder it had at the time! How radiant and comet-like it seemed, and how quickly it has been forgotten and allowed to fall into disrepute. For this reason, I have decided to attempt a short description of this fabulous journey. . .

—Hermann Hesse, *The Journey to the East*

```
┌─────────────────────────────────────────────────────────────────┐
│ ⊠ messages Starting Fresh ▓▓▓▓▓▓▓▓▓▓▓▓▓▓▓▓▓▓▓▓▓▓▓▓▓▓▓▓▓▓      回 │
├───┬─────────────────────────────────────────┬───────────────────┤
│   │  To:                                    │   Will Keep Copy   │
│ ▄ │  Subject:                               │    Will Clear      │
│   │  CC:                                    │    Will Hide       │
│   │                                         │    Won't Sign      │
│   │                                         │     Reset          │
├───┼─────────────────────────────────────────┴───────────────────┤
│   │                                                              │
│   │        ┌────────────────────────────────────────┐           │
│   │        │   Submission failed, reason unknown.    │           │
│   │        │  ┌──────────────────────────────────┐  │           │
│   │        │  │    Send automatic bug report     │  │           │
│   │        │  ├──────────────────────────────────┤  │           │
│   │        │  │ Send automatic bug report & dump core │         │
│   │        │  ├──────────────────────────────────┤  │           │
│   │        │  │        Quit the program          │  │           │
│   │        │  ├──────────────────────────────────┤  │           │
│   │        │  │███████████ Continue █████████████│  │           │
│   │        │  └──────────────────────────────────┘  │           │
│   │        └────────────────────────────────────────┘           │
│   │                                                              │
│   │                                                              │
├───┴──────────────────────────────────────────────────────────────┤
│ Error: Error 0 (in unknown location in unknown location)          │
└──────────────────────────────────────────────────────────────────┘
```

This mail-sending program, a part of the Andrew Message System, goes to great lengths to try to explain to the user the causes of any failed operations. In this case, however, the program has failed in its attempt to determine why its previous operation failed, with results that offer little comfort to the poor user.

Chapter 1

The Hostile Beast

We men of today are insatiably curious about ourselves and desperately in need of reassurance. Beneath our boisterous self-confidence is fear—a growing fear of the future we are in the process of creating.

—Loren Eiseley

We have all heard, far too many times, about the depth, breadth, and profound importance of the computer revolution. Computers, we are told, will soon change almost every aspect of human life. These changes are generally perceived as being for the better, although there is occasional disagreement. But, for better or worse, most of us need only to look around to see proof of the computer revolution throughout our daily lives.

The computer revolution has happened in two stages. The early stage created computers and introduced them to a specialized elite. The second stage, which began in the 1970s with the introduction of the microcomputer, has brought them into our daily lives. Unfortunately, the problems created by the first part of the revolution were (and still are) so immense that many of the researchers who lived through it have scarcely noticed the second wave, in which computers met the common man.

The earliest computers were tyrants—cruel and fussy beasts that demanded, as their due, near-worship from the humans who dealt with them. They indulged in a bacchanalia of electric power, and demanded banquets of flawless stacks of punched cards served up by white-coated technicians before they would deign to answer the questions posed by the puny humans.

As it turned out, most people never joined the cult, and were never initiated into the esoteric mysteries of computation. Indeed, to this day, most human beings on planet Earth have no idea what a GOTO statement is, much less why it might be considered evil or even harmful. Yet this was the topic of the debate of the century for the computer elite—a controversy that still simmers after nearly two decades (Dijkstra 1968 and 1987).

As economic factors have encouraged the spread of computer applications from the laboratory to the office and the home, it has forced the computer to adopt a more accommodating posture. The governing notion today is that computers are tools, objects that can be used by ordinary people for their own ends. The resulting emphasis on simple, flexible, or (misguidedly) "friendly" user interfaces to computers has placed new and unprecedented demands on software engineers.

Imagine, for example, telling a construction engineer that his new bridge had to be "flexible" enough to meet the needs of its users. Perhaps to accommodate people who fear one type of bridge but not another, the bridge might have to be capable of changing instantly from a suspension bridge into a truss bridge, or vice versa. The idea is ludicrous only because of the physical limitations involved in bridges, which are of course subject to the constraints of physical laws.

For better or worse, software is free of nearly all such fundamental limitations. Bound only by logic, software can perform amazing feats of self-modification, of customization to the needs of individual users, and much more. But the cost of such efforts is often very high, especially when it produces (as it usually seems to) poorly structured software systems with high maintenance costs.

The solution of such problems is properly the province of a discipline that has come to be known as *software engineering*. Software engineering is young by the standards of the larger engineering community, having its origins in a conference in West Germany in 1968 (Naur and Randell 1968). But in the computer world, 1968 is ancient history, and the established software engineering practices seem increasingly irrelevant to the reality of user-centered computing. Software engineering has struggled so valiantly, and so single-mindedly, with the incredible problems of creating large software systems that it has for the most part failed to acknowledge the new problems that have been introduced by the demand for better user-interface technology.

The rules of the software engineering game have changed, and unfortunately they have gotten even harder. In particular, the reality of computing today is an increased focus on user interfaces, an area in which the hard-won lessons of the last twenty years of software engineering research are not merely inadequate, but may in some cases actually create more trouble than they can prevent or cure.

Unfortunately, this book does not offer any panaceas, any more than classical software engineering has been able to do so (Brooks 1987). Instead, it seeks to clarify the nature of the problem, and to offer some tentative steps in the direction of a solution. Some of these steps are controversial from the perspective of classical software engineering, although many are established dogma from other perspectives. It should be made clear from the outset that although this book sets forth various claims, it makes no pretense of proving them. The evidence that would definitively confirm or refute these claims does not, for the most part, exist. It is my hope that this book will help to stimulate further discussion and careful, rigorous experimentation.

It should also be understood that most of the particular claims in this book are not new, but have been around in various forms for many years. What I have attempted to do, rather, is to bring together some of the bits of knowledge to be found in various noncommunicating academic disciplines, and to organize these bits as a more coherent vision of how human-centered software can be regularly and reliably built.

What Software Engineering Is For

From a researcher's perspective, it might seem that the last thing software engineering needs in the 1990s is a revolution against orthodoxy. Revolutions disrupt established bodies of theory and practical knowledge, both of which are in remarkably short supply in the world of software engineering. There are really only a few aspects of software engineering that are yet well-enough entrenched to be considered orthodoxy. To understand why one would pick on this poor orphan of engineering, one must first consider the goals and achievements of software engineering as it enters the 1990s.

The basic goal of the discipline is a simple one: to make software better. "Better" is in this case usually expanded into a list of more concrete goals, many of them closely related, some in perpetual conflict. Software is better if it is more efficient, more reliable, less expensive, more easily maintained, more easily used, more easily transported into other environments, and so on. Each of these concrete goals may itself be expanded into conflicting goals. "More efficient" may refer to time efficiency (speed) or space efficiency (memory use). "More reliable" may refer to lack of bugs or completeness of design. Software engineering was born amid a widely perceived but ill-defined "software crisis," and the lack of clarity about the nature of the crisis is reflected in the lack of a coherent goal for the discipline that seeks to solve the crisis.

Such confusion is not, in general, fatal. Engineering is, in a very real sense, the science of intelligent trade-offs. Bridge designers strive toward totally reliable bridges, but they also try to make them as inexpensive as possible, and the two goals inevitably conflict at times. An intelligent and conscientous bridge designer knows, in general, how to resolve such trade-offs, and can rely on a centuries-old accumulation of experience, knowledge, and science. Still it will help, as we proceed to look at what software engineering has accomplished, to bear in mind the fundamental confusion of its goals.

What Software Engineering Has and Has Not Achieved

After quite a few false starts, software engineering is not without significant accomplishments. The staples of the discipline are formal design methodologies and improved development tools. Its impact is clearest in the way the largest of software projects are built, and in the good and bad aspects of the quintessential 1980s programming language, Ada.

Large software projects are the bread and butter of software engineering. Any programmer worth his salt can find some way to keep a horrible five-thousand-line program running; the real test comes with programs of five hundred thousand or a few million lines of code. It is here that software engineering has focused its efforts.

The quick summary of the news on large programs is unglamorous, but useful: you can't cut any corners, and you have to plan everything in advance. This is the basic message behind the plethora of software engineering methodologies that have come and gone in recent years. Indeed, there is evidence, both formal and anecdotal, to suggest that quite a few of these methodologies have, when applied to large projects, yielded significant benefits in the form of more reliable and maintainable software. The resulting programs are still huge, expensive, and nowhere near reliable enough to support the requirements of, for example, the Strategic Defense Initiative (Parnas 1985), but they still tend to look good in comparison to large projects constructed *without* the benefit of such methodologies.

Where small- and medium-scale software projects are involved, the answer is less clear. Anecdotally, there appears to be a dramatic rift between the software engineering "fanatics," who are rumored to write detailed requirements specifications, design specifications, and implementation specfications for a simple bubble-sort subroutine, and the "freethinkers" who consider all such formalism to be nothing more than bureaucratic red tape and obstructionist nonsense. Experimental results have not really substantiated the benefits of applying rigorous formalism to smaller projects, but this could simply reflect the difficulty of obtaining statistically significant results given the smaller dimensions of the projects themselves.

Formal methods notwithstanding, it is probably in the area of tools that software engineering has made the most headway. Software engineers have pioneered the development of programming *environments*—integrated systems to support programmers and make their efforts more productive and coherent. In addition, software engineering has left its mark indelibly on the most important new programming language of recent years, the Ada language. Ada, indeed, is a veritable smorgasbord of software engineering delights. It has elaborate features to support machine independence, including such exotica as compile-time inquiries into a machine's arithmetic precision. With its separation of modules into package specifications and implementations, it takes modular code development to its logical conclusion. Unusual features such as tasks and generic procedures add significant resources to the programmer's toolbox. Finally, by establishing specifications for program development *environments* in tandem with the specifications for the language itself, the Ada designers established the importance of such environments once and for all in the minds of many who were previously nonbelievers.[1]

---

[1]On the negative side, however, many believe that the only thing Ada got right with regard to environments was requiring that they exist, and many also consider Ada to be overwhelmingly complex and feature-laden. A debate on Ada's relative merits is, however, vastly beyond the scope of this book.

Software engineering has accomplished far more than can be surveyed here, but even this brief summary should make it clear that the discipline has made significant contributions. It has not, alas, come at all close to achieving its ultimate goals. Today the "software crisis" is as serious as ever, although we do seem to have gotten rather used to it.

When an effort begins to design a major new user interface, though, managers still get ulcers. The two most predictable results of such projects seem to be a reasonably reliable piece of software that users hate, or well-liked software that nobody understands well enough to maintain. On the subject of why this happens, software engineers are conspicuously silent. Meanwhile, many of those who fancy themselves user-interface specialists are rarely even interested in the question, maintaining that software *engineering* is essentially irrelevant to the *art* of user-interface design.

Why Humans Make Software Messy

From the perspective of the software builder, things are a lot easier if you can keep people out of the picture. Software engineers know—often from years of experience in persuading huge IBM mainframes to perform incredibly complex tasks—the right way to build software. The right way, it turns out, is to design everything in advance, write out careful specifications for every step of the process, and subject those specification documents to review by a small army of your peers before you write a single line of code. Although this description may sound facetious, it is essentially the way things are currently done in large software projects, and it is, by and large, a very good thing that they are done that way, for one very simple reason: it works.

Yes, one can get software to work using a less structured approach, but it generally won't work as well or as long or be as easy to maintain. Moreover, the engineering approach produces software that is almost guaranteed to be more *stable*: given the mountains of specification documents, it is relatively easy to deflect fundamental criticisms of running software with replies like, "you should have pointed that out when we were discussing the design specs." Because the specifications met with approval from all the interested parties, all those parties are reasonably likely to be happy with the way the resulting software works, assuming it does work at all. At the very least, they may feel a bit embarassed about complaining at too late a stage, since this will reflect badly on their own role in the earlier design process.

But this is precisely where user interfaces begin to mess things up. User-interface software should, as its primary mission, make the computer easy and pleasant for humans to use. In most cases, the target humans are relatively unsophisticated, and certainly not a part of the design team. Thus you can't, for example, wave them away by telling them they should have complained when the design specifications were being circulated, because they probably don't even know what design specifications *are*.

This wouldn't be a problem if designers were, in general, good at anticipating what users will want their interfaces to be like. General experience tends to indicate, however, that they aren't. A number of explanations have been offered for the remarkable frequency with which user-interface designers misjudge the wants and needs of their audiences. Jim Morris, the first director of the Andrew Project, has suggested that the problem may lie in the type of people designing user interfaces. The highly rational "left-brained" computer programmers may tend to produce interfaces that the more intuitive, artistic, "right-brained" people despise. This would be an encouraging answer, because we could then let "artistic" people design programs while the more rational folks write them. But there is little evidence that this would work. The Andrew Project (to be described shortly), for example, experimented substantially with consultants from such diverse academic departments as art, design, and English. Excellent results were achieved, but generally only after several iterations—the artists were not, in general, any better able than the programmers to design things "right" the first time around, but they were particularly adept at finding flaws in the prototypes once they were built.

A more discouraging answer, but perhaps a more accurate one, is that people are fundamentally unpredictable. Even the best designers frequently have to make massive changes to satisfy their users. If it were possible to predict human taste in software user interfaces reliably, it would probably be equally possible to predict tastes in music, literature, and new television shows. As modern marketing research has revealed, this is not an easy thing to do.

But if no amount of preimplementation design reviews is going to guarantee the production of a good, well-liked user interface, where does that leave the designers of large systems? For many of them, the answer is intuitively clear. "I wouldn't touch user-interface programming with a ten-foot pole" is a position frequently found among professional software engineers, who can then happily feel superior about the clean, successful systems they build, which happen not to have substantial user-interface components. Naturally, this doesn't endear the software engineers to the self-styled "software artists" who currently do a great deal of the world's successful user-interface design.

The Quagmire of User-Interface Design

Software engineers and software artists generally share the same set of job titles, with names like "systems analyst" or "software specialist," but can readily be distinguished by their approach, and by the kinds of projects in which they excel. The software artist sees the computer as the clay from which to mold a thing of beauty and usefulness. But the medium is far superior to clay in its fluidity. Potters can't continually evaluate how their pots are being used and replace them almost daily with improved versions. The creations of the software artist are never finished, but, in the best of cases, evolve asymptotically toward a stable, graceful state. From the perspective of such artists, software engineering is a cold, sterile science that knows nothing about human beings. Indeed, the idea of firmly specifying an interface before anyone can even try it out is already considered almost barbaric, not only by software artists, but by the community of scientists doing research on human-computer interaction.

Unfortunately, this general disdain on the part of user-interface specialists for the methods and techniques of software engineering leaves them with no solution to the basic problems of software engineering. One frequently finds, therefore, that user interfaces are among the world's most horrible programs in terms of their internal structure and maintainability. When a program is torn apart and rebuilt frequently in response to users' needs, debris quickly accumulates until even the artist stares confusedly at the code written only a few months ago, trying to figure out what on earth it is doing.

Perhaps the best example of this phenomenon is the Apple

Macintosh and the large family of software that runs on it. The Macintosh has a reputation, in large part deserved but frequently overstated, for being the friendliest and most usable machine around. At least, this is its reputation in the user community. In the human-computer interaction research community, feelings are more mixed. Yes, the interface is generally good, but many specific aspects of the interface could probably be improved. It would seem to be an ideal test bed for controlled experiments to evaluate specific improvements, but for one problem: it is remarkably hard to make even the most minor improvements to Macintosh interface.

This leads us to consider the way the Macintosh is perceived by the professional programming community, and notably by the software engineers. "What a horrible operating system!" is a common refrain in such circles. Indeed, some have suggested, tongue in cheek, that one reason there aren't more attempts to copy (or "clone") the Macintosh is that no self-respecting engineer would do it. Despite the widespread acclaim for the Macintosh interface, there hasn't been so much as a whisper of considering the Macintosh as the basis for an operating system standard; proposed standards have instead been based on such systems as UNIX and MS-DOS, despite their obvious absence of any redeeming user-interface qualities.

If the Macintosh experience is any guide, it would seem that producing a good user interface is almost guaranteed to produce the kind of software that software engineering is supposed to prevent. Is it any wonder that the two communities—those concerned with building good software, and those concerned with building good user interfaces—communicate almost not at all?

About the only principle on which the two sides would readily agree is that the user interface should be separated, as much as posible, from the underlying application code, so that at least the latter can be well engineered. This is a starting point, but it doesn't help ensure that user interfaces are built well from both the user's and the engineer's perspective.

A Plan of Action

It is my belief that software engineering and user-interface design, though currently separated by a wide gulf of ignorance and misconceptions, are not fundamentally incompatible. Within the two disciplines, one can find the seeds of a future synthesis of what might be called human-oriented software engineering. Such a science will not spring up overnight, but will emerge slowly from a number of apparently unrelated developments that have been taking place in recent years, in a wide variety of academic disciplines and practical settings.

In order to see the emerging outline of this science, we will approach the problem in three main parts, which, after this introduction, comprise the bulk of the book, and might be viewed as either an attempt to connect several unrelated fields of knowledge or a survey of the current state of the art of human-oriented software engineering. The rest of the book, then, follows this plan:

> *Survey the scene.* Developments relevant to human-oriented software engineering are taking place in a wide variety of fields. Part two of this book informally outlines these developments, their potential contributions, and their blind spots.
>
> *Summarize the folklore.* While a few people have actually done research in the aforementioned variety of fields, a larger number of people have been down in the trenches, actually building user interfaces. Despite the unsystematic nature of their work, a small body of folklore has arisen. Part three of this book presents a sampling of that folklore, heavily weighted by personal experiences in user-interface design.
>
> *Outline the agenda.* The final part of this book, part four, proposes what might be a preliminary agenda for a new discipline of human-oriented software engineering. In particular, it emphasizes how previously unrelated research areas can enrich one another, and how future attempts to build user interfaces could benefit from a recognition of what has been learned.

A Note about the Andrew Project

This book is filled with references to something called the Andrew Project. This is a natural consequence of the fact that I spent over four years working on the user-interface component of that project. In fact, much of this book is a reflection of lessons learned from the Andrew experience. Given this relationship, it is probably useful, at this point, to explain the Andrew Project briefly for the benefit of readers unfamiliar with it.

The Andrew Project (Morris 1986 and 1988) was a collaborative effort of IBM and Carnegie Mellon University. The goal of the Andrew Project was to build a forward-looking software and hardware environment for university computing (i.e., for the needs of academic and research use).

As the project evolved, it produced three main parts, which have seen widespread use:

1. *The Andrew File System* (Howard 1988) is a distributed network file system designed to provide the illusion of a uniform central UNIX file system for a very large network (ten thousand workstations was the design goal).
2. *The Andrew Toolkit* (Borenstein 1990; Palay 1988) is a window-system-independent programming library to support the development of user-interface software. It supports a number of applications, including a multimedia editor that allows seamless editing of text, pictures, various kinds of drawings, spreadsheets, equations, music, animations, and more.
3. *The Andrew Message System* (Borenstein 1988 and 1991) is a very large-scale electronic mail and bulletin board system. It supports the exchange of messages that include the "multimedia" objects supported by the Toolkit, while also supporting "old-fashioned" text-only communication with the rest of the electronic mail world.

Andrew was developed for both eventual wide distribution and essentially immediate use by the Carnegie Mellon campus. The quick exposure of the software to the campus provided rapid feedback for the developers, and thus played a crucial role in the evolution of

Andrew. It also played a crucial role in shaping the thoughts and opinions expressed in this book, which is why so many examples from Andrew are featured here.

Part Two

The Dark Night of the Soul:
The State of the Art in
User-Interface Design

The first part, which is that of sense, is comparable to the beginning of night, the point at which things begin to fade from sight.   And the second part, which is faith, is comparable to midnight, which is total darkness.   And the third part is like the close of night, which is God, the which part is now near to the light of day.

—St. John of the Cross

This night darkens the spirit but only to illuminate it afterward with respect to all things.   This night humbles the spirit, renders it without joy, but only to raise it up and exalt it.

—Daniel Berrigan

To: advisor
Subject:
CC:

What did you mean by 'advisor'?

Advisor

None of the Above

The name 'advisor' is ambiguous. (12:13:28 PM)

Some questions are very hard to answer. In this case, the Andrew Message System software that interprets the names of mail recipients has somehow decided that the address "advisor" is ambiguous, even though it only offers the user a single possible interpretation of the name.

Chapter 2

Who Are All These People?

Science makes major contributions to minor needs.

—Oliver Wendell Holmes

The computer world is well accustomed to sudden changes, to overnight successes, to "hot" new developments that seem to arise out of nowhere. Yet even in this fast-paced environment, the sudden blossoming of research and development in the field known as HCI (Human-Computer Interaction) has been astonishing. By 1988, the premier conference in the field, the annual CHI[1] conference, was attracting nearly two thousand participants and nearly two hundred submitted papers—astonishing figures for an academic discipline that did not exist at the beginning of the decade.

Whether or not there is, corresponding to these large numbers of interested people, a similarly large amount of progress being made in the field is more debatable. The research grouped together under the HCI label seems at times to be an almost incoherent jumble of unrelated topics and techinques. HCI boosters choose to interpret this as a symptom of the dynamic, innovative, interdisciplinary nature of the field, while skeptics can't help wondering if there is, indeed, any common thread to hold the field together.

---

[1]The issue of whether to call the field HCI or CHI has simmered for a long time, with some preferring CHI on the grounds that it is a more pronounceable acronym, while others argue passionately for the symbolic importance of putting humans before computers. Like nearly all of the more important issues in the field, there are few prospects for an early settlement to the debate.

In the 1988 CHI conference proceedings, for example, one can find thirty-nine papers on a wide variety of topics, ranging from descriptions of new interface hardware to experimental evaluations of alternative software techniques to thinly disguised "Artificial Intelligence" projects that purport to offer long-term hope of eventual improvements in user interfaces. Indeed, the introductory messages in each of the CHI conference proceedings make it clear that such diversity is considered highly desirable by the conference organizers, who are themselves in large part the leaders in the field.

But the very diversity of the people interested in HCI research is like a flashing red light to the software engineer, warning that there is danger here. Indeed, psychologists may well outnumber computer scientists in the HCI community, and they are joined as well by rhetoricians, graphic designers, sociologists, and many others. Certainly any of these people may be doing interesting research, but few of them even know what it means to build good, reliable, maintainable software, and fewer of those who do know are worrying about it amid the excited tumult of a newborn interdisciplinary science. Even most of the computer scientists are not really software engineers, but rather researchers pursuing a narrow agenda, independent of the goals of software engineering. In short, software engineers are conspicuously absent from the "interdisciplinary" world of HCI.

Now, although the presence of software engineers would not by itself guarantee that HCI research would produce better structured, more reliable, and more maintainable software, their absence certainly makes it less likely that the multifaceted HCI community will pay any attention at all to the issues that traditionally concern software engineers. Most HCI researchers would probably agree with a statement to the effect that "until we know how to build the right interfaces, there isn't much point in working on really getting the software to be ideally structured and internally documented." In other words, the software engineers aren't, in general, missed at all. The party is going on quite well without them. If the software engineers can be faulted for not coming to the party, then the HCI community can likewise be faulted for not explicitly inviting them, nor even noticing their absence.

Their absence is, however, lamentable, and one of the goals of this book is to stimulate their participation in the future. In this chapter, however, a more immediate goal is to figure out who *is* currently doing HCI research, and what kinds of things they're learning and doing.

As a rough cut, we can divide the HCI community into eight groups, each pursuing a rather independent research agenda that fits more-or-less comfortably under the HCI umbrella:

Technological Experimenters. These are the people who conduct rigorous, controlled experiments to try to determine definitively which, if any, of a small set of user-interface alternatives is preferable, and to test such hypotheses as may arise from theoretical speculations. The goals of these researchers are typically very narrow and limited, and they have been known to achieve solid, scientifically respectable results. A classic example of this kind of work, the editor evaluation experiments of Roberts and Moran (Roberts and Moran 1983), will be discussed in chapter 3.

Sociological Experimenters. These researchers seek to document the effects of the introduction of computing technology to humans (typically large organizations) through careful studies. Such research includes studies of the way that computer technology changes human group dynamics, and has spawned an active new subfield known as Computer-Supported Cooperative Work, or CSCW. Taken together, these first two groups of experimenters constitute the "hard science" core of the HCI effort, and are discussed in chapter 3.

Anthropological Observers. These are the researchers who observe people rather than measure them. They report, inevitably somewhat subjectively, on what humans go through when they learn, use, and become dependent on computing technology. They typically regard their studies as exercises in data collection, in the hope that their reports will prove useful in the formulation of new theories and the design of new experiments. Anthropological observers range in approach from the formal and scientific (examples of which are to be found under the heading of "case studies" in nearly all CHI conference proceedings) to the casual memoir (Kidder 1981; Lundstrom 1987). Although their reports are often clothed in more theoretical garb to satisfy the general requirements of academic publication, their

greatest contribution is often anecdotal, as will be
discussed in chapter 4.

Psychological Model Builders. These HCI "theoreticians"
formulate theories about specific aspects of how the
human mind works, and then build software that will
either test or demonstrate these theories. Often their
work is found under the rubric of "user modeling,"
reflecting the general goal of creating software that
maintains a model of the human user and consults that
model to interpret user input and to formulate responses
(Card, Moran, and Newell 1983). The model builders
are intimately tied to the experimenters, and the two are
discussed together in chapter 3.

Hardware Hackers. These people are trying to build the
future with their bare hands. They typically build
radical new devices for input or output, and report on the
usefulness of those devices for specific applications
(Pearson and Weisner 1986). As their talents most often
lie in building the devices, they frequently devote
inadequate effort to polishing the applications and
conducting experiments to prove their value.

Software Hackers. These are the people who are trying to
contribute to the HCI effort simply by building better
interfaces. Often their efforts focus on tools or systems
to *support* user-interface designs, but at least as often
they simply pursue new ideas in software systems and
report on the results. Since most programmers tend to
like the code they have just written, such reports can be
very hard to interpret objectively. The role of "simple"
innovation, in both hardware and software, will be
considered in chapter 5.

Software Visionaries. These are the people who tend to
wax lyrical on the way the software systems of the
future will revolutionize the way people live their daily
lives. They can be distinguished from the software
hackers largely by the number of adjectives in their
papers, the number of new terms they invent, and the
amount of genuine working software they produce. In
the best cases, they can also be distinguished by the fact
that their visions actually stimulate other hackers to
produce more useful systems.

AI Visionaries. These people are not primarily motivated
by HCI considerations at all, but see the growing interest

in user interfaces as yet another opportunity to promote their vision of artificial intelligence. In the extreme, their papers report on such oddities as natural-language understanding systems that would, they imply, be great general-purpose user interfaces if only they didn't take ten minutes to parse each simple question (Wilensky 1984). Since their claims are generally untestable and their techniques are evidently not yet ready for real user interfaces (the need for fast response is one of the few generally accepted facts of the field), their contribution, if it exists, will not be considered in this book. (This is not intended to denigrate the value of basic artificial intelligence research as such, but merely to place it firmly *outside* the genuine HCI endeavor at the present time.)

HCI researchers clearly pride themselves on the rich diversity of research in the field. There is, they like to think, representation from every relevant branch of science. It seems safe to guess, however, that virtually none of the systems described in papers at the CHI conferences would, when viewed as source code, please the inspecting eye of the software engineer. And conspicuously absent from the CHI publications are software engineering–style titles such as "methodologies for designing and developing user interfaces" or "the user interface life cycle." The absence of software engineering buzzwords is but a symptom of the deeper problem—an absence of software engineering awareness in the HCI community, which cannot be solved by the current HCI researchers alone.

In the next few chapters, we will discuss the HCI research that *is* being conducted in greater detail. Only with a clearer picture of what is being done and what has been learned can we begin to guess how software engineers would fit in as a new component of the HCI community.

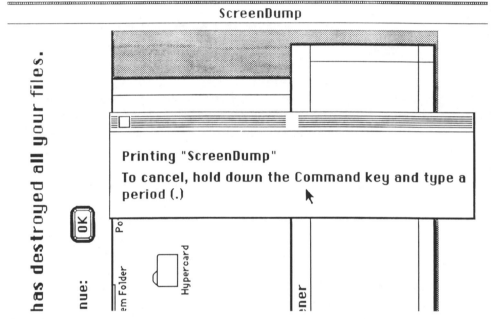

Edit   Eval   Tools   Windows

ScreenDump

has destroyed all your files.

nue:

Printing "ScreenDump"

To cancel, hold down the Command key and type a period (.)

OK

Po

em Folder

Hypercard

ener

How can you get an Apple Macintosh application to produce PostScript output?   You pretend you're actually going to print something on a LaserWriter, even to the point of saying "OK" to start the printing.   Then you quickly—very quickly—hold down the command key and press K. If you're not quick enough, the document will get printed instead of written into a PostScript file.   Of course, the screen you're looking at gives no indication of this magic hidden command.   Remember that the Macintosh is, as of 1991, widely considered to be the world's most usable computer.   This process has reportedly been improved for Version Seven of the Macintosh software.   (suggested by Nat Howard)

Chapter 3

Stopwatches, Videotapes, and Human Nature

Science is simply common sense at its best—that is, rigidly accurate in observation, and merciless to fallacy in logic.

—T. H. Huxley

Science is nothing but developed perception, interpreted intent, common sense rounded out and minutely articulated.

—Santayana

How would one expect classically trained scientists to approach the fundamental questions of human-computer interaction? Undoubtedly, a real scientist would establish goals, formulate hypotheses, and test those hypotheses experimentally. The goals, in this case, are already clear, although multifarious—software should be maximally easy to learn; permit the desired operations to be performed in minimal time; and meet with maximal user approval, as measured by some kind of user-satisfaction survey. The field seems simple, and ripe for application of the classic scientific method.

This is, in fact, the thinking of a substantial portion of the HCI community. Their efforts revolve around theoretical speculation and controlled experiments. Software engineering—in this case, the attempt to improve the quality of the software that implements the interface rather than the quality of the interface itself—is simply not on the agenda of these scientists.

This does not mean, however, that their work can be simply ignored by those interested in the software engineering aspects of user-interface design. There are three major ways in which the "hard science'' work on human-computer interaction can substantially affect the engineering of interfaces. First, and most obviously, such research can prove or suggest that the engineers are frequently doing things wrong. Second, misunderstandings and exaggerations about such research can distort real engineering projects. Finally, software ''research'' projects have a disturbing habit, when successful, of escaping from the lab into the real world, and thus sometimes contribute directly to the proliferation of unmaintainable user-interface software.

The Hard Science of Human-Computer Interaction

Empirical HCI researchers are, by and large, determined to build a hard science base that will be able, at least in principle, to answer most questions about user interfaces definitively. As is customary in such sciences, the work is proceeding along a twofold path of theory and experimentation.

The theoretical research has focused on constructing models of how human users do what they do. In the definitive work to date on the subject, Card, Moran, and Newell (1983) present a remarkably complete broad-brush outline of such a model. They postulate a relatively simple internal computing architecture for the typical human mind, and use this model to predict such things as the time a user will take to complete a given task on a computer. Such predictions are based on both empirically derived values, such as the average time it takes the hand to move from the mouse to the keyboard, and on theoretically derived values for mental response time, which are derived by fitting the data from various experiments to the theory. (To put it more precisely, the theory makes assumptions about the time required for mental activities based on the results of a large set of experiments reported in the literature. While such calculations do nothing to prove the validity of the basic mental architecture being postulated, they do allow the creation of a consistent predictive model, which is no small achievement.)

The natural outgrowth of such theories is experiments to test theoretical hypotheses. It is one thing to build a theory that appears to account for the data from prior experiments. It is quite another to use that theory to predict the quantitative results of substantially different experiments. Experience to date has been mixed in this area, with the typical result being an apparent correlation between theory and reality, but not a strong enough one to make anyone very happy. The generally accepted explanation for reality's divergence from theory is the use of oversimplified models of the user's mind.

It is natural for the actual user-interface designer to ask, regarding such theories, "What's in it for me?" The answer, in this case, is "some occasional rough estimates, but only in a very few cases." The most practically advanced part of theoretical user-interface work is in models like the *keystroke-level model* of Card, Moran, and Newell (1983). These models look at user interfaces at the grain of very simple, basic interactions like keystrokes and mouse movements.

Such a model, for example, would have no trouble predicting the hypothetical but likely experimental result that a user can execute the f keystroke in less time than it takes to execute the control-x–control-v key sequence. Although that particular result isn't terribly useful, such theories can also be used to perform rough calculations of the basic "efficiency" of alternative models of user interfaces. (Here "efficiency" refers primarily to optimizing physical movements; if one user interface forces a user to think longer before acting, this will probably *not* be reflected in the calculations.)

Although the models are not yet precise enough to detect minor differences, they can nonetheless be used to detect large ones. If the calculations reveal, for example, that one interface would require ten seconds for a user to perform a task and another would require thirty seconds, this might be a good reason to prefer the first interface. Thus the theory can indeed be used to help settle a few very specific types of arguments, which is reason enough to expect software engineers to read Card, Moran, and Newell (1983).

In addition to experiments aimed at establishing or disproving theories, another common kind of experimentation in the field is the careful comparison of existing interfaces. In such experiments, two or more "similar" programs are compared in classically structured controlled experiments, with people carefully timed in their use of one or several of the programs. In the broader studies, the desired result might be a statement that "program X is easier to learn than program Y," while in more circumscribed studies the goal might be a statement that "program X is easier to learn with feature Z than without it." A widespread difficulty in this kind of work is getting statistically significant results, as the experiments typically measure relatively brief activities and have to cope with a wide diversity of subjects.

An excellent example of this species of research is Teresa Roberts's work (Roberts 1979; Roberts and Moran 1983; Borenstein 1985b) comparing the time it takes novices to learn to use various text editors, and the time it takes experts to perform routine tasks with such editors. There are many text editors in the world, varying widely in their interfaces but all sharing a large core of functionality. Given these characteristics, Roberts was able to devise a set of experiments that provided useful quantitative comparisons of the ease of learning, and ease of use, of various text editors.

The results of Roberts's work are instructive for both what they did and what they didn't show. The results showed that one set of text editors was clearly better, with a high level of statistical confidence, than another set. However, the results could make no definitive comparisons of the editors within each of these two sets. As it turns out, the better editors were, by and large, the ones that encouraged the user to use a mouse as a pointing device. Thus the experiments contributed substantial support to something that everybody was coming to believe anyway—namely, that the mouse is a good thing—but were unable to contribute substantially beyond this to the "which is the best editor?" debate.

In general, such comparative research seems to be most likely to produce statistically meaningful results only when applied to the biggest differences. But the biggest differences are precisely those that are most likely to be obvious without conducting any experiments in the first place. Until more significant results for smaller differences become easier to obtain, the practical value of such research is likely to remain small.

The unfortunate bottom line, then, is that both theoretical and experimental research in HCI have resulted in a few discoveries of minor relevance to the software engineer building user interfaces, but cannot, in general, offer anything that will substantially influence most aspects of user-interface design. It offers very little guidance to the specific decisions involved in building a user interface, although what guidance it does offer (e.g., "use a mouse" or "avoid long, complex keystroke sequences") is offered with a high level of confidence, and should be heeded.

The Danger of Misconception

If HCI research doesn't offer the practical user-interface designer a great deal of useful guidance, why should the designer pay attention to such research at all? Certainly such vital discoveries as "mice are good" will eventually reach the ears of an interface designer who is completely ignoring the research community. However, the biggest danger in remaining ignorant about basic HCI research is not that you'll miss something, but that you'll hear something inaccurate and reach incorrect conclusions based on a misunderstanding of the latest research.

Indeed, the previous section's brief and oversimplified description of some basic results of HCI research is a perfect example of a dangerously incomplete and partial explanation. For example, from a brief reading of that summary, one might conclude that in the design of a text editor, key bindings are largely irrelevant so long as you have a mouse. In fact, most mouse-oriented editors, including those studied by Roberts, have carefully selected key bindings as well. Similarly, one might erroneously interpret a quick summary of the keystroke-level model as implying that "think time" is unimportant and safe to ignore in performance calculations. Such an incorrect interpretation could lead to an interface with a single keystroke for every conceivable action, although those keystrokes would be so hard to remember that the time it takes users to choose the right one would be enormous.

If for no other reason, then, the professional builder of user interfaces should try to keep up with the latest results from basic HCI research, and in particular to learn about new user-interface ideas and techniques as they emerge, in order to avoid being misled by erroneous summaries or rumors about them.

The Danger of Laboratory Escapes

In biological research, escape from the laboratory is a serious danger, and strenuous measures are taken to protect against the accidental release, for example, of mutated or genetically engineered viruses into the world at large. In software development laboratories, however, the situation is curiously inverted. Nearly every researcher cherishes the secret or not-so-secret fantasy that his or her creation will escape from the lab and be widely used in the world at large. Even more curiously, these fantasies are not frowned upon by managers or regulators, but are benignly tolerated and often even encouraged, albeit with the realization that relatively few such software escapees ever manage to establish a safe habitat in the outside world.

In the cases where such lab escapes are taken seriously and encouraged more systematically, there is even a positive-sounding name for the phenomenon: "technology transfer." At its best, technology transfer is essentially the transfer of ideas: researchers tell developers what the software ended up looking like, and the developers can then reimplement the clever ideas in a stable environment. All too often, however, what really happens is that entrepreneurs try to move the software as directly as possible from the lab to the workplace, with little regard for the software's internal coherence or maintainability.

From the researcher's perspective, any analogy with mutant viruses may seem patently absurd. Surely there is no harm in releasing a useful piece of software to the world. This attitude is particularly prevalent among user-interface designers, who are keenly aware of the terribly primitive user interfaces in common use. What harm can there be in giving the world better interfaces?

The answer, from a software engineering perspective, is almost obvious: the danger is that people will come to rely on software that, because of the basic way it is organized and structured and the peculiarities that have accumulated as it has evolved, is increasingly unreliable and unmaintainable. When software is built for research, reliability is rarely taken seriously, and reliability is notoriously hard to add in after the fact. It makes a good deal of sense, therefore, to treat the notion of software escaping from the laboratory as being a genuine danger, much like the accidental escape of Africanized bees into the Americas. (As with the bees, the public's perception of the danger, once aroused, will no doubt be greatly exaggerated.) Public fears have forced this kind of caution with regard to new technologies such as recombinant DNA research, but have not (yet) focused on the ill effects of imprudent software releases.

This notion is likely to be profoundly disturbing to those who develop experimental user interfaces and cherish fantasies of seeing them used in the real world. In order to address the problem realistically, it will be necessary to propose some kind of development model whereby the ideas developed in the lab can find their way quickly into reliable products, without the widespread release of flashy but unreliable software in between. For this to happen, user-interface designers will have to work much more closely with software engineers. The interface designers will have to accept the importance of seeing their ideas reimplemented reliably, and the engineers will have to accept the importance of picking up new developments from the designers and getting them "out the door" within a reasonable time frame. Neither of these is an easy task.

| Messages | Version 1.24 | jr |

```
ext.cs    (No new messages)
ext.cs.film   (No new messages)
ext.cs.general   (Updates)
116 .Foster (181)
117 r.Husain (920)
118  Finholt  (131)
119 an.Thompson (205)
120 c..      Dale.Amon (414)
121   (85)
122 n.Subrahmanian (328)
123 aniel.Duchamp (74)
124 s        Scott Fahlman  (582)
125  (396)
126 stein (151)
127 (191)
128 Wallstein (671)
129 llen.Walker (78)
130  (265)
131 y.H.Miller (118)
```

on Austria   J.S.4Cruising Sailors?   Rights on: silver raexplorers club   Ge

Folks,

Due to the Electrical Shutdown this weekend, please power off your workstations. This can be done by typing "/etc/halt" in your typescript window. You will then see a message "syncing disks...done". At this point, you can turn the power switches on both the monitor and CPU unit to the "off" position.

Thank You,

Marybeth Cyganik

Error: Not owner (in open in GetPartialFile)

Cruising Sailors? A spectacularly broken user interface, unretouched. A programmer can look at a picture like this and laugh ruefully, but what can one expect the user to do when confronted with a dialog box such as this one? The most common response is raw fear and panic.

Chapter 4

That Reminds Me of the Time . . .

Memory believes before knowing remembers.   Believes
longer than recollects, longer than knowing even wonders.

—William Faulkner

The first program of mine that other people depended on was a library
automation system.   Among other things, the system ran a nightly
batch job that printed out catalog pages in need of update.

I had recently fixed a bug in which a user-oriented part of the
software died a horrible, messy death when it ran out of disk space.
By adding a few lines of code to the program that died, I ensured that
in the future, if it ran out of disk space, it would pause, print an
informative message, wait until space became available, and then
continue. I prided myself on a user-friendly and robust solution.

In fact, this solved my problem so nicely that I decided to
practice a little preventive maintenance, and added this safety feature
to *all* the programs in the system.   All my basic system tests still
worked, so I installed the new version of the software and went home
to bed.

The next morning, the computer center was a morgue. The only
activity to be seen was the professional staff, huddled in confusion
around the console. "What's up?" I asked.

"We're out of disk space," came the harried reply, "but every
time we manage to free up a little space, it seems to get filled again
within seconds. I don't know what could be doing it."

I felt a cold chill that has since become all too familiar: is my
software making dozens of people miserable? Am I about to look

like a total idiot? As casually as possible, I logged in and killed the batch job that I knew in my heart was causing the problem. Instantly the system sprang back to life, and though I tried to slip away quietly, everyone immediately figured out whose fault the problem was.

As it turned out, there had been a long-standing bug in my batch job. At the end of each night's run, when all the new catalog pages had been printed, an infinite loop filled up the entire disk. Previously, when that happened, the program would die, the temporary files would be automatically deleted, and (since the job ran in the dead of night) nobody ever noticed a problem. My clever new fix, however, had ensured that whenever the disk filled up, my program would simply go to sleep and wait for more space. My "preventive maintenance" had crippled the campus computing facilities for several hours. Fixing software has never seemed quite so simple to me since.

The Joy of Anecdotes

Every professional programmer has a collection of stories such as that one, most of them rueful tales of insidious bugs. Many of these stories are highly instructional, and might be thought of as the Aesop's Fables of software engineering. Like other fables, these stories often have explicit morals, but often, too, their value can't be neatly captured by simple slogans.

Nowhere is this more true than in the world of user interfaces. As a graphic example, the picture at the beginning of this chapter shows a genuine, unforged example of a user interface gone truly berserk. Some question had been scheduled to be asked in a dialog box, but the entire screen display is corrupted, giving evidence of serious memory corruption, and the dialog box delivers the coup de grace by apparently propositioning the user for inscrutable, but probably unsavory, purposes. But instead of selecting "yes" or "no" responses to this solicitation, the user is invited to choose one or the other of a pair of Rorschach blots. It is instructive to look at that picture reflectively and ask how many different things could be wrong with this program, and to speculate on what the "wrong" answer might cause to happen. Which option would *you* choose, if you were the user? How can you even begin to diagnose the problem, or to program in such a way as to avoid it in the future?

What Does It All Mean?

Stories and pictures like these are among the most fun aspects of being a programmer. They are traded, among professional programmers, like old war stories were traded in previous generations. Often they are filled with insights, with meaning, and with cautionary words for younger programmers. They are unscientific, unsystematic, and not included in any part of the standard curricula of software engineering or computer science. Are they, in fact, good for anything other than passing time in a bar? What is the real value and role of anecdotes?

The intuitive answer that most programmers would give is that such stories are indeed useful. This is the same sentiment that produces the widely heard refrain, "I learned more in my first six months on a programming job than I did in four years of working toward a degree in computer science." The essence of programming anecdotes is a distillation of practical programming wisdom, the very entity that formal courses in computing completely fail to supply.

Of course, one person's educational anecdote is another's boring old shaggy dog story. For every person who reads a great novel and declares fervently that it "changed my life," there is one who reads the same book and finds it inscrutable or self-evident. But just as most would agree that novels are, on the whole, of value, in large part because they can affect people so profoundly, so, too, are anecdotes of general value to programmers even if certain particular anecdotes leave us cold. Stories that other programmers have seen fit to tell about interesting programming adventures are fun, educational, and unwelcome in almost any part of the current curriculum.

Inasmuch as programming is not now, nor likely soon to become, an exact science, it is not too much of an exaggeration to claim that these anecdotes are, in fact, the fundamental storehouse of our knowledge about the craft of computer programming, or at least the raw data for that knowledge. If this seems absurd to the academic mind, we must recall that practicing programmers routinely declaim the sheer irrelevance and uselessness of nearly everything that is commonly taught under the rubric of computer science, except for the actual programming courses. Down in the trenches, the soldiers have thrown away the weapons they were given, and are choosing instead to fight their battles bare-handed. The folklore emerging from these battles is of value both in its own right and in trying to figure out why the weapons were discarded in the first place.

Electronic Anthropology

If anecdotes are useful, the bottom line remains that it is not at all clear what, in particular, they are useful for. Computer science and software engineering are very young fields of study, and are still taught in the styles of mathematics and electrical engineering, the disciplines from which they evolved. Neither of these disciplines has felt much need for the kind of knowledge derived from anecdotes, so there is no place for them, no established framework for fitting the anecdotal knowledge into the field as a whole. The few examples in the literature in which researchers have tried to present lessons from a set of anecdotes, such as Orr's report on Xerox's experience with field service on copiers (Orr 1986), have had to clothe the anecdotes in a great deal of additional material in order to make them acceptable for publication in a given context.

However, the situation is by no means unprecedented in the sciences. Anthropology often operates in a very similar situation. Much of the most useful information in that field has been derived from firsthand accounts of different cultures by observers and even by members of those cultures. These firsthand accounts are by no means all there is to anthropology, but they are the fundamental data that inform the more theoretical and analytical work.

Indeed, anthropology is an excellent model for studying the process by which software is built, for another reason: the most interesting part of this process is the most human part of the process, the design and use of the software itself. Inasmuch as human-oriented software engineering is the study of this process, it could be argued that it is more properly a branch of anthropology than of the mathematical sciences. The study of software creation may, in fact, be grossly misclassified in the academic world today, leading to a distorted overemphasis on formal models and a lack of basic work in collecting the raw data that comes, most often, in anecdotal (or at least nonquantitative) form.

Software engineering would not be the first discipline to borrow heavily from anthropology. Ethnomusicologists run all over the world collecting native music from a wide variety of cultures, basically operating on faith that saving and preserving this "raw data" will eventually prove valuable. Indeed, the music thus collected has led to insights in music theory, and has also inspired modern composers in varied and striking ways. So, too, the collection of careful observations and folk legends about programming might pay dividends in unpredictable ways.

An Informal Debugging Technique

Tracy Kidder, in *The Soul of a New Machine* (Kidder 1981), tells the story of how the last major hardware bug was found in a new Data General machine. As the deadline for bringing the machine to market approached, the designers were frustrated by their long-standing inability to find a "flakey" bug in the ALU (Arithmetic Logic Unit) of Gallifrey, their prototype machine. A flakey bug is one that the debuggers can't figure out how to reproduce. In this case, the designers had spent days trying to catch the bug when it occurred, but had been stymied by its unpredictability, its apparently randomness.

> On October 6 the vice president, Carl Carman, came down to the lab as usual, and they told him about the flakey.
> The ALU was sitting outside Gallifrey's frame, on the extender. Gallifrey was running a low-level program. Carman said, "Hmmm." He walked over to the computer and, to the engineers' horror, he grasped the ALU board by its edges and shook it. At that instant, Gallifrey failed.
> They knew where the problem lay now. Guyer and Holberger and Rasala spent most of the next day replacing all the sockets that held the chips in the center of the ALU, and when they finished, the flakey was gone for good.
> "Carman did it," said Holberger. "He got it to fail by beating it up."

In this case, the designers were so used to finding bugs in their design that something as basic as a loose chip never crossed their minds. How many others have lost days of effort by failing to question the reliability of the underlying subsystem? Is there any better way to make that point than by telling this story?

User interfaces have come a long way since the days of TECO, an infamously powerful but hard-to-use line editor that was widely used on DEC machines for decades. TECO horror stories are so abundant that it was hard to choose just one example. This one was supplied by Craig Knoblock:

"As you probably know, TECO is a text editor in which all of the commands are control characters. To enter some text you would type control-a, followed by the text, and a control-d to end the text. When I was first learning TECO I decided to type in a ten-page paper. I typed control-a, followed by all ten pages of text, followed by the control-d. Unfortunately, as I was typing in the paper I must have hit another control character. So when I typed in the final control-d I received the message: 'Unknown control character—input ignored.' An hour of typing down the drain."

Chapter 5

The Quest for the Perfect Line Editor

In the long run men hit only what they aim at. Therefore, though they should fail immediately, they had better aim at something high.

—Henry David Thoreau

Having invited a pompous business acquaintance to his private club, the businessman greeted his guest and offered him a drink.

"No thank you," the guest replied, "I don't drink—I tried it once and didn't like it."

Later, inviting his guest to join him in a game of cards, he received the same reply: "No thank you, I don't play cards—I tried it once and didn't like it."

An invitation to play billiards elicited a similar reaction, followed by an awkward pause. At length the guest continued, "of course my son Charles is really quite skilled at billiards, and plays frequently."

"Ah," said the host, "your *only* son, I presume?"

—as told by Debbie Borenstein

The computer world is probably about as close as any human activity will ever come to Mao Tse-tung's dream of "permanent revolution." Revolutions big and small have become the norm, at least for the first few decades of the computer age.

In the late 1960s, a few developers pioneered a new type of software called a *screen editor* (Stallman 1981), which became widely available by the end of the 1970s. The basic new idea of screen editors was to use most or all of a video screen to show the user a relatively large portion of a file at one time—typically twenty-three or twenty-four lines. As the user changed the text, the screen display was constantly updated to reflect those changes. This was indisputably a revolution in a world where ''line editors''—editors that would only allow changes to be made to one line at a time, and didn't even keep an updated version of that line constantly visible on the screen—were the only editors most people had. Line editors were the best that people could do with the previous generation of terminal hardware, largely teletypes and other paper-based displays. In hindsight, it seems clear that screen editors were a natural, or even obvious, outgrowth of the improvements in terminal hardware, particularly the introduction of the video screen.

Calling such developments ''obvious'' is, however, a gross disservice to the people who invented them. In particular, it is worth noting that even in the 1980s people were still building line editors, and other people were in fact still arguing over what the ''ideal'' line editor would look like. (This argument probably still persists even today, but it has at least moved out of my range of hearing.) Given the indisputable superiority of screen editors to line editors for most tasks, such arguments have become entirely irrelevant, except inasmuch as their persistence is instructive for studies like this one.

Why did the line editor die such a lingering, painful death? Why did a great number of intelligent people waste time building newer and better line editors long after screen editors (and indeed even the subsequent generation of editors, mouse-oriented large-screen bitmap display editors) had become available? And perhaps most troubling, why do some people still prefer to use their favorite line editors twenty years after they have become unarguably obsolete?

The Introduction of New Technology

Even in the constantly changing computer business, new technologies are born into an extremely hostile world. For various reasons, this seems to be particularly true of user interfaces. The developer may emerge wild-eyed from his office, ranting to anyone who will listen about the breakthrough he has made, about how easy his new gadget or program is to use, about how it will revolutionize the way people talk to computers. His colleagues will smile uneasily and shift restlessly from one foot to the other, pondering several alternatives, all of them unpleasant:

> **The developer** has finally gone over the edge, lost his mind, and will no longer be useful for anything.
>
> **The developer** has reinvented the wheel, and will eventually have to face the disappointment of knowing that his technique isn't new at all.
>
> **The developer** has indeed made a minor innovation, but nobody is really likely to care about it because it is irrelevant for most purposes. (This is probably the *best* alternative the colleague can think of.)
>
> **The developer** is telling the truth, but the world is not ready for his breakthrough, and it will languish without aggressive and expensive marketing. This will be painful for both the developer and the observer.
>
> **The developer** is telling the truth, and the world is about to sit up and take notice. The colleague is going to have to take the time it takes to learn how to *use* this new breakthrough. Since he is not (yet) aware of any need for the breakthrough, this is an unappealing prospect.

Thus the first hurdle that any new interface technology faces is in getting anyone at all to take it seriously. True revolutions are hard to understand at first, and people need to be convinced that it is worth the bother of trying.

Assuming that enough people have been convinced to make it clear that there is more to the innovation than a few lunatic programmers with an obsession, the second hurdle to be faced is *practicality*. Quite a few innovations are, in fact, impractical in the real world, but many more of them are initially thought to be impractical by upper management. Throughout the industry, one can find research labs full of truly astonishing user-interface hardware and software, yet it was considered a breakthrough when Apple actually made a *mouse* and a window-oriented user interface an integral part of a commercially marketed computer. Both mice and window managers had by that time been floating around in research labs for over a decade, but had not been taken seriously by the people who sold the computers. An impressive number of innovations are now similarly waiting for their day. It should be noted, however, that this particular hurdle is much harder for hardware than for software, because software can be test-marketed and hence "proven" more easily by entrepreneurs.

Assuming the practicality hurdle is overcome, and a new technology makes it to market, the final and toughest hurdle must be faced: human inertia. Although a few people play with computers for the sheer fun of it, most use them as tools, to get their work done. Such people see each new innovation as another annoying detail they may have to master before they can get back to their real work. This is why some people have used line editors for more than a decade after better alternatives have become available to them. If you force people to use new technology, they'll resent the coercion, and if you don't force them, they may never use it at all. Technological "revolutions" are simply too frequent for most people's taste, and resistance is perfectly natural, especially when the costs of learning a new technology come close to outweighing the benefits.

Naturally all of this is a great frustration to people who devote their own working lives to trying to improve the human-computer interface. Is there anything that can reduce the barriers blocking such improvements? Is there a shortcut by which innovations can travel more quickly from laboratory to workplace?

Hardware Innovations

In user interfaces, the toughest kind of innovation to get people to accept is hardware innovation, primarily because hardware is generally useless without software. Thus, it is almost inevitable that after months of work on your new hardware, you will still have a virtually useless piece of equipment to sell to people.

When a new hardware technology is brought to the attention of potential users, they are usually smart enough to ask the obvious question: *What is it good for?* Unfortunately, the hardware designer (or the hardware's advocate, who is rarely the actual designer once the case has progressed a bit further) is most often so excited about the hardware, so intuitively certain that it will be spectacularly useful, that he is essentially unprepared for this question.

People treasure their work desks as their own private worlds. Both the person with the compulsively neat desk and the person with the cluttered fire hazard share this in common: their desks reflect both their ways of working and their images of themselves as workers. It is only the fact that there are amazingly useful programs such as word processors and spreadsheets that has convinced many people to accept the presence of a computer on their desks. A similar resistance to externally imposed changes to the desktop may be reflected in the suspicion with which some people view the mouse. "Why do I need to have one of those things taking up space on my desk?" is a common and telling question.

This psychological barrier is the one that keeps virtually every potentially useful innovation in user-interface hardware from reaching the general user community. There are dozens of such technologies that could be used today, notably moles (foot pedals) and data gloves (gloves that track every movement of a hand with impressive subtlety). Interesting though these technologies are, and exciting though they may be to their admirers, they are exceedingly unlikely to come into widespread use without the motivating force of a useful and desirable piece of software that depends on them. Of course, there is a chicken-and-egg problem here; it is hard to get anyone to invest a lot of time, money, and energy in designing software for extremely specialized or "nonstandard" hardware. As a result, most demonstrations of new hardware show some kind of existing software modified to make use of the new hardware. For example, moles are often demonstrated by replacing the mouse driver in a text editor with a mole driver. The obvious reaction to such a demonstration is, "Cute, but I've already got a mouse and can do whatever I need with that, and the area under my desk is useful storage space."

There are several ways to fight back against such reactions. In recent years, the advocates of moles have conducted studies of the general efficiency of using moles for text editors; the basic idea is that by replacing the mouse with a mole, you allow the user's hands to stay on the keyboard, and thus eliminate the time it takes to move hands from keyboard to mouse and back. In an effort to document this claim, they have conducted controlled studies of the use of moles for editing tasks (Pearson and Weisner 1986).

Such research is good science, and may have convinced a few people to think seriously about moles. However, it hasn't gotten very many people excited. How much time, after all, do people really spend moving their hands back and forth from keyboard to mouse? The answer, of course, is that in total the amount of time spent this way is quite large, but it is rarely enough in any one activity to be a noticeable nuisance. Thus there is no perceived value, for most users, in trading in mice for moles, and there are evident drawbacks: the trade would require a reorganization of the desk and a retraining period to get used to the new hardware.

To get people excited enough to move moles out of the labs and into the office, then, will take something more than dry studies showing seven-percent gains in efficiency. Something *exciting* has to happen. Someone will have to develop an application in which the value of moles will be so overwhelming that everyone will want one. Preferably such an application would use moles in a way that mice could not be used, perhaps as a third input device to be used in conjunction with keyboard *and* mouse. Perhaps it would exploit a familiar physical analogue, such as the pedals in a car, making the application seem more ''natural'' to the user. Unfortunately, mole advocates have not been able to think of such a watershed application.

But it is, at any rate, not my goal here to rescue the mole from obscurity; quite possibly it belongs there. My point, rather, is that something was missing from the work that developed the mole—namely, a clear idea of what the thing was actually *for*. If developers seriously hope to create hardware that will move from the lab to the workplace, this question should be on the table from the first day of work. If there isn't a clear answer, it might be reasonable to ask why the hardware project is even being funded. If there is a clear purpose to the hardware, however, a substantial effort will be required to build the software that demonstrates that purpose. In general, the selling of new hardware technology is going to take a substantial software effort, and it doesn't hurt to begin that effort even before the hardware has actually become available.

Once the software is built, however, the combined software and hardware package faces the same hurdles that any new innovative software must face, even if doesn't use any unusual hardware. These hurdles are quite substantial themselves, which explains why new hardware technology is hardest to sell: the hardware developer has to sell both the hardware and the software that demonstrates it, and usually isn't even aware of this need. Nor, of course, is he aware of the hurdles that any such software *alone* must face. For example, to this very day, text editors for the IBM PC are largely doomed if they are entirely dependent on a mouse, because the mouse is not standard equipment and is absent from many PCs. This makes it more difficult to build a successful text editor that does a good job exploiting the mouse when it *is* available.

Software Innovations

Innovative software is seductively easy to get people to use. If it is truly innovative, it usually isn't too long before a few people get genuinely excited about it. Real innovations like spreadsheets and lightweight processes have taken their natural user communities by storm, genuinely revolutionizing them overnight.

But while such software seems to take wing of its own accord, it only rarely reaches a very high cruising altitude. Once the core set of "natural enthusiasts" has found the software, a plateau is often reached. For the great majority of users, software is simply not something one *ever* gets excited about. Even if the software is genuinely and obviously useful, many people resist it to the end.

The bottom line is that software is, for the masses, only a marginally useful thing. For each user, a few pieces of software are incredibly important, but the rest is fluff. Most users know this intuitively. They also know, therefore, that any new software that comes along is most likely to be fluff as well. Moreover, this is not merely stubbornness on the users' part, but simple common sense. The proponent of a new program should always remember two things upon meeting resistance from such users: first, the users' skepticism is well justified by the frequency with which new software is largely irrelevant to their needs; and second, the skepticism may be well justified even in the current case. That is, if you tell users that they need your software and they remain skeptical, there's a good chance they're right. They know what they need far better than you do.

How can we tell whether slow user acceptance of new software is indicative of the stubbornness of the users or of the marginality of the software? There's no easy way. The only really useful thing to do is to ask questions of the skeptical users: *Why don't you use the new program? Have you tried it? Did you find it difficult to use? Did you find it useful?* The answers to such questions, if asked in a nonthreatening manner, can be extremely illuminating. Often it reveals the "tried it once and didn't like it" syndrome, in which one ugly or cryptic feature was enough to make the user give up entirely on the software. Such revelations can lead to "minor" improvements in the software that make it possible for future new users to avoid the ugliness, and thus possibly to become regular users of the program.

In a relatively small or geographically compact potential user community, this tuning process can also sometimes lure back the people who "tried it once and didn't like it" as well, but one cannot, in general, count on such second chances. Often a person who is turned off by software once will simply never run it again at all.

Engineering for Change

People love stability. People rightly expect that next year's car models will still have the gas pedal on the right and the brake on the left. If you proved that reversing them had some major benefits, people still wouldn't want any changes. Such is human nature.

Computer technology, on the other hand, doesn't seem to admit even the notion of stability at present. (Witness the lack of standard computer keyboards, even within a single manufacturer's product line.) It is a whirlwind of change, prompting worried articles and symposia on such subjects as "managing change" and "lifelong education." How, such articles ask, can we control the pace and the nature of change to make it more palatable to people? Put another way, they ask how people can learn to live happily with constant change.

One might as well ask how we can learn to love the bomb. A shift in perspective would have us asking, instead, how we might *minimize* and *hide* change to the greatest possible extent. While this may seem to conflict with what we know of technology, it is much better psychology, if it can be done. One of the best ways to get people to switch text editors is to pretend the new editor is just an improved version of the editor they know and love.

Calling a complete rewrite an "updated version" is neither unheard of nor dishonest from the perspective of the user. All that users really care about are the commands they have to issue and the reliability and speed with which those commands cause work to get done. If the internals of a program are rewritten, that's fine with the users.

For example, the Andrew Project built a program called EZ, a text editor strongly reminiscent of Emacs, the editor most popular with Andrew's target user community. It can be argued that the designers of EZ made a tactical error by not giving it a more Emacs-like name. Had it been named something like "Emacs-plus" or "Seemacs," for example, it might have encountered somewhat less resistance from die-hard Emacs users. As it was, many Emacs users didn't realize that EZ was very much like what they were used to, but with a host of fancy enhancements and a few important omissions.

But what about when the interface *does* change, due to some kind of genuine innovation? In such cases, the lie is harder to pull off. If, however, the previous interface had a clearly defined and consistent "style" of interaction, it will probably be possible to make some substantial changes and maintain the users' belief that they are using "the same program." Judging which changes can and can't be made, however, is exceedingly tricky.

Perhaps the best bet, in the long run, is to design programs explicitly to facilitate user-interface evolution. One technique is to provide several alternative user-interface "flavors." The user of a given piece of software might be given three user-interface flavors to choose from, which would differ substantially enough to tell them apart and to attract different kinds of users to each. A crucial consequence of providing such choices is that the users come to think of the program as *more than just its interface.* They are thus less likely to perceive a new release as being "a different program." But the main benefit of the "flavors" approach to user interfaces is that it provides a graceful evolutionary path for the program's actual user interface. Major changes can be introduced as new alternative interfaces, while old ones can be supported for transitional periods. New releases can include announcements like "Release 2.3 is the last release that will support the Butter Pecan user-interface flavor. Butter Pecan users are encouraged to begin using another flavor, and in particular may wish to try the Butter Cashew interface that was introduced in Release 2.1." In this way, users at least have control over the timing of changes to their interfaces. They are not forced to learn a new set of commands just when a deadline is pressing, but can select a time when they have a few hours to figure out the new interface. Most users are sophisticated enough to appreciate, in such a situation, that the software developers have made a substantial effort to ease the pain of the transitition.

A final, serendipitious advantage to the "flavors" approach is that it encourages a very clear division between a program's functionality and its user interface. This is a good idea for a whole host of reasons, most notably the mercurial nature of user-interface code. If software provides multiple user interfaces from the start, it is hard to avoid making the code separation clean and complete.

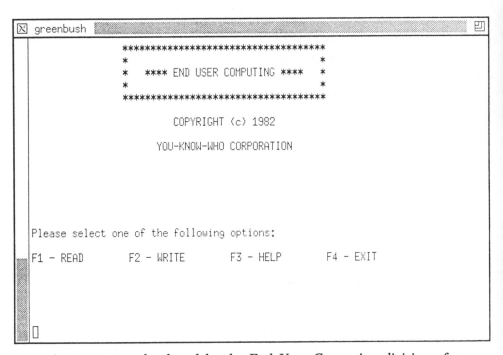

```
 ┌─────────────────────────────────────────────────────────────────┐
 │ ⊠ greenbush ░░░░░░░░░░░░░░░░░░░░░░░░░░░░░░░░░░░░░░░░░░░░      ⊡ │
 ├─────────────────────────────────────────────────────────────────┤
 │         ************************************                     │
 │         *                                  *                     │
 │         *    **** END USER COMPUTING ****  *                     │
 │         *                                  *                     │
 │         ************************************                     │
 │                                                                  │
 │                    COPYRIGHT (c) 1982                            │
 │                                                                  │
 │                 YOU-KNOW-WHO CORPORATION                         │
 │                                                                  │
 │                                                                  │
 │                                                                  │
 │  Please select one of the following options:                    │
 │                                                                  │
 │  F1 - READ       F2 - WRITE       F3 - HELP       F4 - EXIT      │
 │                                                                  │
 │                                                                  │
 │  []                                                              │
 └─────────────────────────────────────────────────────────────────┘
```

A system was developed by the End User Computing division of a European company.   During user testing, one subject stopped working after the initial screen display appeared. Why? "It told me to stop," said the user, pointing to the large, unambiguous words at the top of the screen. One has to wonder why the words were there in the first place. (suggested by Ralph Hill)

Chapter 6

The Men in Suits

"If the law supposes that," said Mr. Bumble, "the law is an ass, an idiot."

—Charles Dickens, *Oliver Twist*

Laws too gentle are seldom obeyed; too severe, seldom executed.

—Benjamin Franklin

Software is built by people. User-interface software is influenced by a wide range of people. The users, of course, have a major influence on the design of good user-oriented software. The personalities and tastes of the designers and implementers inevitably play a big role in shaping the user interface. Scholarly researchers conducting controlled experiments and propounding theories about the human-computer interface have their influence as well. Still, there is one group of people that has not yet been mentioned in this book, but that has a major impact on the quality and style of the final interface—and indeed on nearly every aspect of software development. These people go by several names, but are known most commonly to the programmer as "the men in suits."[1]

The men in suits include, typically, lawyers and executives, both within the company developing the software and outside it.

---

[1] I apologize for the obviously sexist title. In this case, some comfort can be taken from the fact that although the people in suits are nearly always men, that fact speaks rather well of women.

Programmers of all persuasions tend to dismiss them all out of hand in the most outrageous language. "All lawyers are pond scum" is entirely typical of the programmer's rhetoric. Of course, it is always said in a joking tone, but the universality of the sentiment warrants closer scrutiny. How do lawyers and executives affect the process of building software in general, and user interfaces in particular?

Outright Coercion

One of the major skills exercised by lawyers is, of course, coercion. The point of many legal actions is to stop people from doing something they want to do, or to force them to do something they don't want to do. In the case of software, lawyers have overcome some initial disadvantages to reach a position in which they can seriously undermine a rational development process, particularly in the area of user interfaces.

As a historical note, it is fascinating to recall that only a few years ago, the U.S. legal system seemed firmly committed to the notion that software was not patentable. This position stemmed from early legal maneuvers by IBM, which was making all its money from computer hardware and trying to ensure that no competitor could undermine its position by producing unique software that would have the protection of the patent laws. IBM used its formidable staff of lawyers to obtain some very clear rulings about the unpatentability of software, and this was unchallenged for many years. In the business community, however, it was quickly and widely perceived that IBM had made a serious strategic mistake in pursuing this ruling. It seemed that many of IBM's best ideas, which were being copied by others, could have been protected by the very patents that IBM had successfully argued should not be obtainable on software.

From the developer's perspective, however, IBM's mistake was heaven-sent. In the absence of *patent* protection, software was protected only by the *copyright* laws. The copyright laws, however, protect only the actual code that implements a program, not the ideas behind it. Thus a user interface, for example, could be copied with impunity, so long as the copy was a completely new implementation, including no actual source code from the original. That this was a boon to developers does not indicate that developers are natural cheats and copiers, but rather that ideas tend to evolve gradually rather than spring suddenly into existence. A good idea in one developer's code tended to be picked up and refined in the next implementation, so that the ''copy'' more often than not improved on the original.

This situation was a constant source of delight to developers. Contrary to what might have been expected, developers were rarely angry to see their ideas copied and improved in this manner. Instead, it gave them pleasure to see their ideas reaching a wider audience. Often, friendly rivalries would follow, in which two or more developers would produce successive versions of a similar program, each building on, and improving on, the competitor's previous effort. (One of the more famous examples of this is the Emacs text editor, which improved continuously through a series of such versions; see Stallman [1981] and Borenstein and Gosling [1988] for details.)

Less happy with this free flow of ideas, however, were the men in suits. To them, perhaps not surprisingly, stealing was stealing, and the developers who were hiding behind this ''loophole'' in the patent laws were simply thieves. The fact that each new version often brought significant improvements paled beside the fact that each improved version meant less revenue for the owner of the original program.

In the 1980s, therefore, the situation began to change. A few large corporations decided to invest some of their legal resources in setting new precedents to establish the patentability of software. It will be some years before the results of this effort can be fully assessed, but already a cold wind has blown through the programming community.

The Macintosh Menus

The clearest example of this trend is the patent issued to Apple Computers for the style of menus implemented on the Macintosh personal computer. In essence, Apple drew a line in the dirt and dared anyone to cross it. The Macintosh certainly has reasonably nice menus, but Apple's claim to have invented them is dubious, to say the least. In fact, the Macintosh menus are the product of the standard kind of evolution discussed earlier, and are basically an improved version of software that came out of other companies (notably Xerox) and universities.

Few professional programmers would argue that Apple's patent on this menu interface is deserved. Indeed, the patent is widely derided by user-interface developers, although rarely loudly enough to let Apple hear, and most believe that a serious legal challenge could have the patent struck down. Yet the patent has not been challenged or struck down, and several companies are paying royalties to Apple in order to be able to use a similar menu style on their own products. Other development efforts, such as those at universities, where royalty payments would be problematic, make a strong and unnatural effort to distance their menu interfaces from Apple's.

Thus by its patent, and by related copyrights that it also defends aggressively in court, Apple has created a fundamental schism in the evolution of such interfaces. If the patent holds, there will be, in the future, two lines of menu evolution—the menus that are produced by Apple and its licensees, and the menus that are produced by nonlicensees. This in itself will be inefficient, but not a serious problem. The greatest danger will occur if other companies manage to patent other, related aspects of user interfaces. As more and more good ideas come under the protection of patents, it may become increasingly unlikely that any one program can incorporate the state of the art in user-interface design without sinking into a quagmire of unending royalty payments and legal battles. (Indeed, in 1989 the chickens came home to roost at Apple, which was sued by Xerox for, in essence, stealing its user-interface technology. The outcome of the particular lawsuits is hard to predict, but it is easy to predict that the battles will, in the long run, benefit no one so much as the lawyers.)

Fortunately, this seems unlikely to happen in the short term, as several alternative menu interfaces have been developed that are clearly not Apple's and that are at least as usable as Apple's, and probably better. (The Macintosh interface appears particularly well suited to small screens, but grows increasingly annoying as screens get larger.) Apple's benefit from the patent may prove to be short-lived, and the negative implications in programmer hostility and Apple's possible overcommitment to a single menu technology may prove somewhat longer lasting.

An interesting thought to ponder, however, is what might have happened if Apple had taken a more liberal policy toward licensing. Imagine, for example, that the royalty payments on the ''Apple menus'' had been truly small, and nominal or nonexistent for universities and other nonprofit organizations. In that case, Apple's menus might have remained part of the mainstream of menu-interface evolution, and people might still be paying substantial royalties (in small increments) to Apple many years from now.

Blackmail

Software patents may be used as a form of outright coercion, providing protection against theft of ideas at a potentially high cost to future inventors. Beyond such coercion, however, the copyright and patent laws also offer the less objectionable but nonetheless lucrative option of licensing. A specific copyrighted program or program fragment, for example, may be licensed by its owner for use in someone else's products, and the copyright holder will then receive royalties on the sale of that product. When prudently managed, licensing can provide impressive returns to the holder of a copyright or patent.

The UNIX Operating System

AT&T, for example, holds the copyrights and even some patents on the UNIX[2] operating system. This operating system is very widely used, and is now sold by nearly all of AT&T's competitors, who have to pay royalties to AT&T for each copy sold (except for a few UNIX imitations that are completely free of AT&T code).

How did AT&T manage such a feat? How can other companies similarly maneuver themselves into a position where they will receive royalties on their competitors' products? The simple answer is that it was an accident, a mistake. If the men in suits at AT&T had had any idea what was really going on, they would almost certainly have killed it. But the way it happened is nonetheless instructive for those who would like to repeat the "mistake."

---

[2]UNIX is a registered trademark of AT&T UNIX System Laboratories. You can tell from the fact that I put this footnote here that they also protect their hold on the software by using the trademark laws. However, all the trademark laws protect is the word "UNIX," rather than any of the actual source code. The latter is protected by the copyright laws and by a patent on one specific aspect of the system.

Basically, UNIX was developed by a pair of clever developers working in Bell Labs, the research end of AT&T. In its early versions, UNIX was extremely unpolished, a thing only a developer could love. Seeing no commercial value in it, and because of earlier tangles with antitrust laws, AT&T allowed it to be distributed nearly for free to a wide range of universities and research institutions. (It did not, however, place UNIX into the public domain, but rather, on the prudent advice of its lawyers, maintained all legal rights but licensed them for next to nothing.) This placed it in the hands of a lot of other clever developers, who improved it in a wide variety of ways. By the time a decade or so had passed, and the men in suits began to become aware of its possible commercial value, UNIX was the product of many thousands of man-years of effort. A fair amount of that effort was put in by people who didn't even work for AT&T itself, but who, in the spirit of the effort, simply passed their improvements back to AT&T, which made them part of the standard distribution.

AT&T thus found itself in the enviable position of owning all rights to a popular product that it didn't even completely develop in-house. As the demand for UNIX grew, AT&T, sensing a new profit center, began to invest more in internal UNIX development work and hike the licensing fees, as well. Despite the increased fees, the fish was already hooked. UNIX was already a de facto standard in many quarters, and even mighty IBM now pays substantial royalties.

The surest way to destroy AT&T's potential profit from UNIX at an early stage would have been to charge real money for it. In its early days, UNIX included some neat ideas, but if it had cost too much, people would have simply reimplemented it from scratch. Now, even though it has begun to cost real money, it has matured both in internal complexity and in its role as something that users demand, so that a total rewrite is a much more difficult option to contemplate. But still, the cash cow is not guaranteed to be immortal, and AT&T's men in suits could conceivably yet manage to kill it with rapacious licensing fees. Already, a number of groups have developed or are developing "uncontaminated" versions of UNIX, without a single line of AT&T source code. If AT&T starts to charge too much, one of these versions could conceivably replace AT&T's as the "standard" from which future systems will evolve.

The Andrew Window Manager

An interesting contrast to the UNIX success story is the less well known but far more typical tale of how the institutionalized greed of the men in suits managed to kill another promising piece of software, the Andrew[3] Window Manager. In contrast to the UNIX story, which occupies a key role in the history of computer software, the Andrew Window Manager is nothing more than a footnote in that history, a minor story that has been quietly repeated many times without anyone ever seeming to learn anything from it.

The Andrew Window Manager (WM) is a program that, as its name implies, manages windows on a computer's screen. It was one of the first network-oriented window managers to run under the UNIX operating system on a scientific workstation with a bitmap display. It was fast, easy to use, and reasonably reliable. Among those who used this class of machine, it generated intense interest, and a steady stream of visitors came to its birthplace, Carnegie Mellon University (CMU), to see it.

However, WM was not owned by CMU. It was developed as part of the Andrew Project, a joint venture of IBM and CMU. Part of the agreement that defined the joint venture stated that IBM would own the software, but that "reasonable" licensing arrangements would be available.

Unfortunately, there are many definitions of "reasonable." To a university, a licensing arrangement like the standard UNIX license was "reasonable." To IBM, such low-cost licensing sounded insane. While IBM and CMU argued over licensing arrangements, the people waiting for licenses got impatient.

One such group, from MIT, eventually gave up on WM entirely, and built their own window system instead. That system, which they called X Windows, had the traditional evolutionary relationship with its predecessor: it did everything WM could do, and more. Moreover, the MIT group managed to align itself with a multivendor consortium that funded the continued development of X Windows as a nonproprietary, easily licensed standard window system. Within a few years, IBM found that nobody even *wanted* to license WM any more, and that IBM was in danger of being entirely left out of an

---

[3]Andrew is a trademark of IBM Corporation. Unlike AT&T, they don't (yet) seem to care whether people say so in footnotes or not.

emerging standard. With little choice, IBM embraced the X Windows standard, and CMU began converting all of the Andrew application software from WM to X.

What is most notable here is that WM was a very promising and useful piece of software. It was ahead of its time, and many groups would have liked to pick it up, use it, and improve it. By trying from the beginning to squeeze every possible penny out of it, IBM squeezed the life out of it instead. Good software needs to evolve, and it cannot evolve in the face of greedy licensing arrangements.

As an epilogue to the story, IBM seems, to its credit, to have learned something from such episodes, and has been making most of the other software produced by the Andrew Project available with extremely liberal licensing arrangements; it remains to be seen whether IBM's men in suits have now found the right balance between protecting their investment in software and facilitating the natural evolution of that software.

Peer Pressure

There is no law that says that all cars' dashboards must look essentially similar. There isn't even a federal law that says that all 110-volt electrical outlets have to be the same size and shape, though local building codes generally require this. In both of these cases, however, there is a strong perception on the part of the manufacturers that *standardization* is a good thing. Having a roughly standard arrangement on a car's dashboard makes it easier for people to figure out how to drive each new car they sit down in, and having a rigorously specified standard for electrical outlets makes nearly every electrical applicance work in nearly every outlet in America.

Businesspeople are, by and large, very comfortable with and comforted by standards. There are few nightmares worse than a product that is incompatible with everything else in the world, and when such products do emerge, they usually die a quick and expensive death in the marketplace. Conforming to a standard helps to increase a product's chances of success, and defying a standard is usually taking a big risk.

Consider, for example, the personal computer market. In the 1970s, each computer vendor brought a radically different machine to the market. Consumers were terrified of buying a machine that would become obsolete, that would never have further software written for it. The need for a standard was obvious and indisputable.

Then IBM introduced the IBM PC. By the sheer force of IBM's market position alone, the PC quickly became the de facto standard for personal computers, despite its painfully obvious technical inadequacies. The importance of a standard was so compelling that consumers and manufacturers alike ignored all technical factors in a headlong stampede to standardize on the PC and its operating system, MS-DOS. In an astoundingly short time, all competitive operating systems and machine architectures virtually vanished from the personal computer marketplace, as each vendor rushed its "IBM clone" to market.

It is especially worthwhile to take note, therefore, of the only personal computer to defy this trend with major success—the Apple Macintosh. The Mac introduced an entirely new computer: its processor was different, its operating system (although scarcely worth dignifying with the term) was different, and even the devices used for input and output were substantially different—it used a bitmap display instead of a character-oriented one, and it relied heavily on a pointing device called a *mouse*.

Although no computer scientist is likely to put forth much effort to defend MS-DOS as an example of what operating systems should be, a defense of the Macintosh operating system is even less likely. Apple had relatively little technical expertise in operating systems when it built the Mac, and this shows everywhere. Yet despite the technical step backward, and in the face of an ongoing rush to standardize on the PC, the Macintosh survived, eventually earning substantial profits for Apple.

The Macintosh prospered for one simple and indisputable reason: it was much easier and more pleasant to use than the PC. Standards can survive all sorts of *technical* inadequacies for the simple reason that nobody who *uses* the product is likely to understand or be bothered by such inadequacies. But noticeable user-level inadequacies can generate significant resistance to standards. If the PC hadn't had a rather poor basic user interface, it might have become a permanent standard, but in the face of the Macintosh challenge, among other factors, even IBM abandoned MS-DOS in favor of a new system, OS/2, intended both to be a better operating system *and* to support a better user interface.

There were two big things wrong with the PC: it had a bad

operating system and a bad user interface. The operating system became a widespread standard, while the user interface was gradually replaced. The clear lesson is that standards are more difficult to apply to user interfaces than to other aspects of software, because people are more likely to notice the effect of a bad standard.

Today, there is widespread sentiment among users that all computers should look like Macintoshes. This is, in part, a reaction against the PC, and also a reflection of the fact that most users are unfamiliar with any third alternative. Fortunately, any effort to standardize on the Macintosh interface is slowed somewhat by the obvious unsuitability of the Mac operating system as a standard.[4] Taking the Macintosh lesson into account, however, one should be skeptical of any attempt to standardize user interfaces. Even where such standards are established, it seems, true improvements are much more likely to destroy the standards than in other areas. If users are ever convinced that a new interface is so much better than the Mac's that it's worth the pain of changing over, no amount of standardization is going to keep a large number of them from wanting that new interface.

Slavery

All employees work for employers; that is one of the basic realities of holding a job rather than being self-employed. The situation of the software developer is no different in this regard from that of many other workers, but the effects of this situation are nonetheless worth considering as another example of how the men in suits shape the way software is built.

To begin with, programmers suffer along with many other workers in facing deadlines that are often unrealistic and arbitrary. In software engineering, overly inflexible deadlines are known for promoting programs that are fragile and hard to maintain due to programmers' cutting too many corners in their haste to meet a deadline. Among user-interface designers, deadlines are generally perceived as cutting off experimentation too early, and forcing the designer to make choices without sufficient experimentation. In both

---

[4]It is also slowed by Apple's predatory licensing policy, as described previously.

cases, the result is the same: a product that is in some sense poorer than what might otherwise have been produced. Of course, managers can't very well tell programmers to "take as long as you like," but the programmers' own estimates of the time required for a project are ignored at the project's peril. (Indeed, Fred Brooks makes it clear, in *The Mythical Man-Month*, that such estimates are notoriously *low*, and that far more time should be scheduled for a project than the programmers think necessary.)

There are often many other seemingly arbitrary requirements imposed on programmers from above. Most notable among these is compatibility with previous products by the same company. Although such requirements often make a great deal of sense from a business perspective, they can have a deeply stultifying effect on the design of the next generation of software. These constraints are thus easily misused. If the older product has a relatively small number of users, the gain of keeping those users happy may not be worth the cost in restricting the new product. If the older product has a large number of users, it might be worth considering completely separating the effort to support the old product from the effort to create a new one; in this way, the old users are kept happy without putting a straitjacket on the developers of the successor product. What is important to remember is that this is not merely a convenience for the developer. The new product is more likely to be successful, and hence the company more likely to make money, if the developers are not overly fenced in by compatibility requirements. (Of course, in the ideal case the original product is so well designed that it is relatively easy to make future products compatible, but this is only rarely the case.)

Finally, the worst kind of arbitrary requirements often imposed by management are those born of sheer ignorance, superstition, fear, or laziness. A manager who learned how to program in the 1960s might tell programmers, "You must write your programs in FORTRAN," simply to avoid having to learn a new language. A manager afraid of being criticized from above for a mistake in hardware selection might opt conservatively for all-IBM equipment, telling programmers they can use nothing else. In each of these cases, the damage done might be major or nonexistent, depending on the real *technical* requirements of the project; if the technical realities are ignored in framing such requirements, any kind of results can ensue.

Intelligent Management Is Not an Oxymoron

The foregoing discussion may seem to imply that software developers, or at least the author of this tirade, are implacably hostile to the people who manage software development, but this not the case.   The hostility is real, but is in large part directed against misguided management practices that, though common, are far from universal.   If the job of the men in suits is primarily to manage the creation of software so that it makes money, there is absolutely no fundamental conflict with the desire of programmers to create a better piece of software.   The scant literature documenting real-world programming projects seems, however, to indicate that the managers who understand how to let software creators do their job are quite rare (Lundstrom 1987).

But if there is not fundamental conflict, there is, however, a great deal of room for misunderstanding the process by which software is created, and for misapplying lessons from the management of other kinds of dissimilar activities.   How, then, can one manage a software development project, particularly one with a large and troublesome user-interface component, without unnecessary interference in the successful development of that project?

Unfortunately, this is a very difficult question to answer.   The optimistic title of this section derives primarily from observation. Some projects *are* in fact managed intelligently, and the programmers working for such intelligent managers often produce programs that are both good and profitable.   Figuring out the secret of such managers is a bit more challenging than observing that they exist.

My own experience suggests that the key factor may be *recognition of reality*.   Too often, managerial dictates seem to stem from a misperception of the technical realities of the programmer's situation.   This underlying problem may be manifest in unrealistic deadlines, overly constraining compatibility requirements, or any number of other constraints.   What is often difficult for the manager to admit is that *the programmer almost inevitably knows the technical reality better than the manager, even though the latter has the final authority to make decisions.*   Therefore, the manager must always be sensitive to his or her own relative ignorance in technical matters. Although the manager may have to push the programmers to the edge of their own estimates, one can only push to a certain limit before reality intervenes.   If a problem is primarily technical, it is usually best to let the decisions be made, as much as possible, by the most technically knowledgeable people.

In user-interface development, the situation is complicated by the fact that even the programmer has a hard time seeing reality clearly. The manager must remain sensitive to the fact that *nobody* might know the best answer in advance, and that substantial experimentation might be required. Conversely, the worst results are likely to come from a manager who prescribes a user interface in detail, as that person may be powerful enough to enforce such requirements yet far enough removed from the development process to fail to notice when the interface just isn't any good. If a problem is primarily one of user interface, it is usually best to let the decisions be made, as much as possible, by iterative design and observation of users—a process from which most managers are far removed. The possible negative role of managers is discussed further in chapter 14.

The GNU foundation, which probably represents the most constructive of programmers' efforts to get back at "the system," distributes the above license with its software, all of which is given away for free. A less constructive example is that of a programmer at Apollo who had several years of work thrown away when it turned out that the computer he was programming wouldn't fit through the hatch of the submarine for which it was designed. The programmer's revenge? Most computers to come out of Apollo since that time have had a special error code built into the system. If that error number ever occurs, the error message to be printed out to the user will be "unit will not fit through 25-inch hatch." (submitted by Murray Spiegel)

Chapter 7

Information Wants to Be Free

A little rebellion now and then is a medicine necessary for
the sound health of government.

—Thomas Jefferson

Managing software development is rather different from managing
most other enterprises and, as discussed in the previous chapter, it is
frequently seriously botched. This has led to a disturbing rift in the
culture of computing between the managers and the managed. Within
the programming community, one can find innumerable programmers
who regard management as either "the enemy" or "the clowns." As
previously stated, this does not happen everywhere; good
management is not an unheard-of thing in the computer business,
merely a rare commodity. Rare enough, in fact, to drive substantial
numbers of programmers to a radically antimanagement position.

That position is taken to its logical extreme by software
anarchists who regard the pirating of software almost as a moral
imperative. Though their position is scarcely defensible, some of
their basic perspectives are, in fact, shared by a large number of
programmers. It seems that programmers are, either by their nature or
as a result of their experiences, strongly predisposed to be skeptical
about the way the business end of the information age has been
organized.

Against the backdrop of a fiercely competitive computer
industry, a new counterculture has quietly emerged. This is an
anarchic culture of people who, though they may call themselves
"hackers," are law-abiding, responsible citizens whose primary

loyalty is not to their company but to an elusive and almost mystical vision of their new art, to the very idea of building *good software*. All over the country they meet—never in any organized manner—to talk about software. If one works for IBM and one works for DEC, that rarely inhibits them from saying or doing anything. Indeed, they will frequently work to get each other access, through legal or illegal means, to software that should, according to the men in suits, be protected at all costs. These are people who, to varying degrees, simply don't play the game that the men in suits are paying them to play.

The game they *are* playing is harder to pin down. Mostly, they are just computer enthusiasts, tired of seeing promising projects destroyed by uncomprehending management. They are by no means in open revolt, and often don't even think they are doing anything that management would disapprove of. They are wrong in that management is usually strongly opposed to the free exchange of information, though they are often right in the broader sense that they aren't really doing anything that will harm their company. The few companies with the vision to encourage such free exchanges have prospered quite nicely. Indeed, the recent industry-wide vogue for "open systems" is a hopeful sign that some, at least, are beginning to see both the futility of trying to keep software secret and the value of sharing it.

A good example of the mainstream of the "anarchist" subculture can be found in the distributed electronic bulletin board facility known as USENET. USENET is a totally informal network of many tens of thousands of computers all over the world, which exchange hundreds of megabytes of information on hundreds of "newsgroups" with topics ranging from computer hardware to nude bathing to Islamic mysticism. All of this takes place on corporate computers, at (relatively low) corporate expense, primarily in the form of network and disk usage. A fair portion of these companies participate in USENET knowingly, recognizing that the benefits of having happy employees and of having quick access to the technical discussions far outweigh the costs of carrying the discussions about New Age music. However, an impressive number of corporations have no idea that they are paying for such activities at all. One participant signed a message with the amusing disclaimer, "The views expressed above are my own, and not those of [his company's name]. Actually, I think I'm the only one here who even knows what

USENET is, or that we're getting it!'' On the various newsgroups, a clear sense of community is easily found—a fraternity of technical people from hundreds of corporations, with a shared distrust of all management.

Management, in the eyes of disgruntled programmers around the world, is often so stupid that it can't even see the clearest way to make the money with which it is so obsessed. Just as a child must learn to let go of one toy in order to grab another, so, too, many managers have yet to learn to permit some of their ideas to get away for free (or nearly free) in order to make money on others.

But, if the hard-core anarchists see the men in suits as irredeemably greedy and shortsighted, and the pragmatists see them as mostly greedy and shortsighted but retaining the potential for enlightenment, there is still a third school of thought that sees the whole system itself as misconceived. These are the people who rally behind such slogans as "Information Wants to Be Free" (Brand 1987). These people want to stop playing the "company X owns program Y" game entirely, and replace it with a completely different approach to the economics of software.

At a grass-roots level, this has manifested itself in the phenomenon known as "shareware." In shareware, software is distributed free, along with permission to copy it and a plea for money. Users are encouraged to try out the software, to see if they like it, and to send money to the author if they find the software to be of value. While nobody seems to have gotten rich on this utopian scheme, some people are making a living at it, and it certainly short-circuits all the fences (copyrights, patents, trademarks, copy-protection schemes) that the men in suits have tried to build around *their* software.

Within the professional programming community, the rejection of the established order has been most clearly set forth by a group known as the Free Software Foundation. This group starts from the premise that nobody should own software. They encourage people to give them software, no strings attached, and they then proceed to improve and redistribute it. Their position is essentially that charging money for software is a form of theft.

It seems likely that almost every programmer in the world would happily go to work for the Free Software Foundation if it weren't for the minor question of money. When all is said and done, it remains an undeniable fact that no matter who pays for software, and no matter how much or how little they pay, the people who write software still have to eat, and still enjoy fast cars and fancy camcorders. In this sense, the universe is structured so that the men in suits have the developers cornered, and they know it.

Perhaps surprisingly, software buyers seem to have even less enthusiasm for free software than do the software developers. Most people really believe, at some level, that you get what you pay for. In the case of free software, this is often untrue, but the general perception is hard to shake, and frustrating for the people trying to give their software away. Certainly it seems unreasonable to expect free software to come with the same level of user support, bug fixes, and documentation that users rightly demand from commercial software.

Still, the adversarial nature of the relationship between programmer and management is a powerful warning that something is amiss. Highly paid programmers in the 1990s should be among the happiest people in the history of mankind. If they aren't, it may be because they are being treated badly, not in material ways, but in what might best be described as spiritual ones. Imagine van Gogh being told, when his masterpiece *Starry Night* was ninety percent complete, that management had decided there was no market for it, the project was being scrapped, and the canvas had already been burned. Would he have been comforted to learn that his job was still secure, and that in fact he was even getting a healthy raise to make him feel better about it? If managers occasionally thought of their programmers as aspiring artists in engineers' clothing, they might be less likely to impose maddening arbitrary restrictions, or to discourage free association with other artists. And the programmers, for their part, might even stop trying to find ways to destroy the very notion of software ownership.

Part Three

The Ten Commandments:
Principles for User-Interface Design

And He wrote them upon two tablets of stone, and gave
them unto me.

—Deuteronomy 5.19

When the state is most corrupt, then the laws are most
multiplied.

—Tacitus

Do what thy manhood bids thee do, from none but self
expect applause.

—Sir Richard Burton

```
 X  typescript ~/writing/hose/figures                                    回
   This message is in 'x-be2' format.
   Do you want to view it using the 'ezview' command (y/n) [y] ? y
   ---Executing: ezview
   Now executing the EZ program -- if an EZ window appears, you can quit it by
      holding down the middle mouse button and selecting the 'Quit' menu.

   You can prevent the window from appearing by INTERRUPTING (CTL-C) now,
      in which case you should see a text-only version of the data.
   Starting ez (Version 7.0, ATK 15.1); please wait...
```
```
           X  ez /tmp/metamail.4099.5153                          回
             Let f(x,α) be a probability density function for x; fα and fαα are its first
             and second partial derivatives. G and U are utility functions. The
             optimization problem is to choose a function s(x) in [c, d+x] and a value
             α so as to maximize


             ∫ G(x − s(x))f(x,α)dx        (1)


           Cancelled.
```

*The Andrew Project devoted an enormous amount of effort, in its
early days, to making its user interface easy for novices to learn. The
pop-up menus, in particular, were the subject of several massive
redesigns, with frequent testing to determine what users found easiest
to learn and use.*

*Unfortunately, the testing methodology always assumed that the
user would be briefly trained in the use of the menus before
attempting to use them, rather than be left to flounder with no help at
all. This minor assumption turned out to have a critical impact on the
design of the Andrew menus. Andrew's menus, it turns out, are very
easy to use if someone first tells you that they pop up when you hold
down the middle button on the mouse. To the new user who is not
told that "secret," the Andrew menus are mystifying in their total
invisibility. The above program, which opens an Andrew window on
the screen of someone who may not be an Andrew user, has to go to
great length to tell the user how to find the Andrew menus and choose
"quit" when he or she is done.*

Chapter 8

Never Underestimate Your Users

I think that we may safely trust a good deal more than we do. We may waive just so much care of ourselves as we honestly bestow elsewhere. Nature is as well adapted to our weakness as to our strength. The incessant anxiety and strain of some is a well-nigh incurable form of disease. We are made to exaggerate the importance of what work we do; and yet how much is not done by us! Or, what if we had been taken sick? How vigilant we are! Determined not to live by faith if we can avoid it; all the day long on the alert, at night we unwillingly say our prayers and commit ourselves to uncertainties. So thoroughly and sincerely are we compelled to live, reverencing our life, and denying the possibility of change. This is the only way, we say; but there are as many ways as there can be drawn radii from one center. All change is a miracle to contemplate, but it is a miracle which is taking place every instant.

—Henry David Thoreau

After spending the first few decades of the Information Age writing software that only an expert can use, programmers have awakened to the fact that the real action is with the common man. The Apple Macintosh, the most notable example of this trend to date, has gotten tremendous mileage out of the fact that its complete primary instruction manual can be read and comprehended by nearly anyone in a few hours. Indeed, that is an impressive milestone in the history of personal computing, and today everyone has leaped headlong into the race to make software easy for anyone to understand.

Unfortunately, this all too often takes the form of programmers "talking down" to the "dumb users." In the minds of many programmers, making software easy to use is necessary only because most people are too unsophisticated to understand the glories of a *real* computer system. When programs are built with that attitude, the resulting product can be downright offensive to intelligent users.

What every programmer must remember is that the average user is likely to be reasonably intelligent, and might even on occasion be smarter than the programmer. However, despite (or perhaps even, unfathomably, *because of*) this intelligence, the user must be assumed not to want to spend a lot of time learning to use the computer. Moreover, he may want to use the program relatively infrequently without having to relearn a lot each time. The target audience, therefore, for most user-interface software, is a clever, sophisticated, but impatient user.

This turns out to be a difficult audience for many programmers to understand. Why, one must wonder, would any intelligent person *not* want to spend the time needed to learn how to use the system in great detail? This question undoubtedly sounds ludicrous to any nonprogrammers reading it, but it underscores a very real prejudice to be found among programmers, particularly young ones. Programming and using computers is, to these people, the most exciting intellectual activity they have ever imagined. It is but a short step from this observation to the conclusion that those who lack a passion for computers must be lacking in intellectual curiousity, drive, or ability.

Such prejudices are common in every field. Artists look down on scientists as cold. Scientists look down on businessmen as greedy. Businessmen look down on artists as out of touch with reality. And, of course, doctors and lawyers tend to look down on everyone. In most cases, these professional prejudices are essentially harmless. But for user-interface programmers, they are a real danger, because the prejudice can prevent them from doing their job properly.

It might be argued, by way of justifying the programmer's prejudices, that the advertising profession has demonstrated the economic value of talking down to people. Most advertising seems to be aimed at an extremely unintelligent audience, and yet seems effective in persuading most of us. But the software industry may, I suspect, provide a few exceptions to the old saying that "nobody ever went broke underestimating the intelligence of the American people."

In bad advertising, we are often subjected to a short burst of inanity, usually too short to make us feel that our intelligence has been genuinely insulted. In using software, however, we may interact with it for hours. The same level of insult will sting much more strongly when it is so prolonged. Imagine sitting down to listen to three solid hours of television commercials for a single product every day. If the advertisers knew you were going to do that, they'd quickly learn to make it much lower-key and respectful of your intellect. But that is precisely what you do with a computer program.

It is vital, therefore, that a computer program be targeted at the correct level of audience. Anyone using a word processor, for example, should be assumed to be fundamentally literate and probably fairly sophisticated regarding the basic ideas of word processors. (If they aren't, they will be after using the program for a few days or weeks.) The word processor must therefore not be offensively hand-holding or patronizing.

More important, it must not omit, in the name of "simplicity," the kinds of features that a sophisticated user will demand. This is a classic programmer's ruse to avoid work. A programmer who says, "that feature would confuse novice users so we shouldn't provide it," may really mean, "I don't feel like implementing that feature." Certainly it is easy to confuse the novice user, but it is also easy to "hide" advanced features so that novices don't need to know about them and won't "trip over them," but experts can get at them if they need them. It is therefore rarely necessary actually to exclude a feature for the protection of novices.

An illuminating example is the early history of text editors in the Andrew Project. The original designers of the Andrew user interface had a great deal of experience with text editors. In particular, that experience centered around the Emacs family of editors, which has achieved great popularity in the research and academic communities. Emacs and its derivatives are extremely powerful and feature-laden editors. One can use Emacs for years and remain ignorant of a vast number of its most advanced features. On the other hand, one can write hundreds of pages of programs using an extension language for Emacs, and indeed whole suites of software have been built "inside" Emacs (Stallman 1981; Borenstein and Gosling 1988).

The Andrew Project, however, was not intended to produce tools for the technical elite, but rather for the masses. The strongest criticism of Emacs, over the years, has been the difficulty that novices often had in learning it, and the general complexity stemming from its richness. Thus, when the time came to design a multifont screen editor for the fancy bitmap displays being used in the Andrew Project, a major retreat was made from this complexity. The first Andrew editor, known as EditText, was deliberately made far less powerful than Emacs or many other text editors. It was, in fact, delightfully easy to learn to use, and experiments showed that new users were happily editing multifont documents with it after only a few minutes of instructions. It seemed, at first blush, a phenomenal success.

Then, gradually, a strange thing started to happen. More and more people seemed to be using Emacs itself (which was available to Andrew users, although it lacked any advanced features for fonts and graphics), and abandoning EditText. C programmers, and hackers in general, never really embraced EditText at all, which didn't surprise anyone. What was more surprising was the way the general population gradually abandoned EditText. Instead of producing multifont documents right there on the screen and then printing them, people more often used Emacs to produce input files for a traditional document production system (such as troff, TeX, or Scribe), which would then produce multifont paper output.

This seemed an enormous step backward, but the reasons were clear enough if you talked to people. EditText just wasn't powerful enough for these "unsophisticated" users. Each of them had found valuable features in Emacs and in their document production systems that EditText simply lacked. And while it was certainly true that they didn't know how to use a large portion of the fancy features of those other systems, that didn't seem to matter very much, so long as they could do what they needed to do. In fact, it seemed that the wealth of yet-unlearned features was a comfort to many, a reassurance that if they ever needed to do something different, there was a good chance it would be supported by their tools.

The EditText program, accordingly, died a slow and lingering death. It remained extremely useful as a first text editor for novices, and thus was not phased out entirely until the subsequent generation of Andrew editors had matured. But the lesson was quite clear for those who were watching: the need to make a program easy to learn absolutely does *not* imply that the program may omit basic useful features. Even the most casual users will quickly learn to spot a tool's glaring inadequacies. The novices who learned EditText so quickly were inexperienced, but they weren't stupid or unsophisticated, and it didn't take them long to realize they had been given a lawn mower's engine in a Porsche's body. Without an option to upgrade to a more powerful engine, they quickly decided to trade in the whole car—in this case, on last year's model, out of date but still running strong.

Even the best of user interfaces can't hide a fundamental lack of functionality. The good interface must be built on top of an adequate base of an underlying program that meets the users' genuine needs.

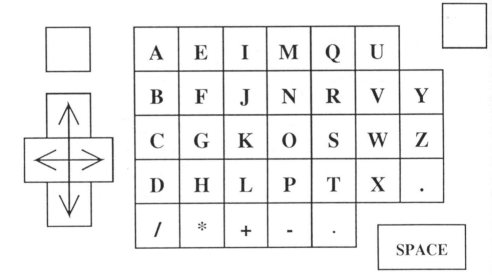

A	E	I	M	Q	U	
B	F	J	N	R	V	Y
C	G	K	O	S	W	Z
D	H	L	P	T	X	.
/	*	+	-	.		

SPACE

The AWACS aircraft represents the state of the art in air defense technology. Aboard the airplane is a host of computers that are used by highly trained technicians to monitor all air traffic. For all the millions of dollars spent on the system, they apparently could do no better in the realm of computer keyboards than the layout pictured above.

Chapter 9

Pretend That Small Is Beautiful, but Don't Believe It

Small is Beautiful.

—E. F. Schumacher

Explanations should be as simple as possible, but no simpler.

—Albert Einstein

For a while in the 1970s, it seemed that a popular slogan of the time, "Small is Beautiful," described some fundamental truth about life itself. Certainly in computer hardware, smaller has always been better, as hardware engineers work ceaselessly to pack more computing power into an increasingly smaller box.

Many experienced users of computers will strongly defend the notion that small is better in software, as well. Bigger, fancier, and more complicated programs, they will tell you, suffer from diminishing or negative returns as new features are added. "This program is simply too complicated," is their common complaint. Undoubtedly larger programs tend to be buggier and less maintainable than small ones, leading many to question the cost-benefit trade-offs in a larger, fancier program.

This perspective is easy to sympathize with. The plight of the casual user is perhaps the best case for smaller software. The casual user of a computer program is someone who uses the program repeatedly, but irregularly and infrequently, perhaps no more than an hour or two a week, but who needs to keep returning to the program

about that often. Such users will most likely *never* become truly
expert in the use of the system. They are, indeed, doing well not to
forget what they know from one week to the next. They frequently
consult the instruction manual or help system for reminders of things
they already knew, and they are rarely interested in learning new
features, since they have trouble remembering what they've learned
so far. Casual users are often very important people, perhaps high-
level managers who don't have the time to use the computer more
often. When such VIPs say that software is too complicated, people
listen.

No programmer, of course, makes programs excessively
complicated on purpose. Behind every feature in a program, there is
a story. Either the feature was considered important when the
program was originally specified, or it was added to the specification
while the program was being built because the need seemed clear, or
it was strongly requested by a user of the program. Rarely is any
feature added to a system without some such clear and apparently
correct rationale. The aggregate effect, however, of many strongly
desirable features is often a program that appears overly complex.

The difficulty is quickly illustrated for anyone who tries to
simplify an existing and well-used program. If simplifying means
eliminating the least important features, this turns out to be a
thankless and usually pointless undertaking. For each feature
eliminated, some segment of the user community—typically *not* the
ones who have been complaining about the program being too
complex—will cry in loud anguish over the loss of some "vital" (to
their purposes) bit of functionality.

This is inevitable because the cries are coming from a user
community that is largely separate from those who complain about
the program's complexity. People who have to use a program every
day, all day, or who at least rely heavily on a program and use it very
frequently, rarely complain (once they've been on the job for a while)
about a program's complexity—even in the case of really bad
software, where such complaints are most justified. They have
internalized that complexity, and know well how to use the program,
however complex it may be. After becoming such experienced users,
they have gradually learned how to exploit many of the very features
that make the program complex. Indeed, the heavy user of a program
often comes to *depend* on its complexity for shortcuts in a wide
variety of situations. From the perspective of the programmer who is

simply trying to make a program as powerful and useful as possible, these expert users are the "best" audience.

Unfortunately, no program can be written for only this single audience of experts. Even programs that are typically only used by such expert users still have to be used by novices, in training to become experts—and at that stage the complexity is still a barrier rather than a set of useful features. Programs that are intended to be used *only* by casual or novice users would clearly do well to sacrifice some features on the altar of simplicity, but most programs play to a mixed audience of widely varying expertise. What is the programmer to do?

Flexibility and Power Scare the Novice

The Emacs text editor, mentioned briefly in the previous chapter, provides an instructive example of a program that casual users often decry as "overly complex." This editor provides nearly every feature ever provided by any nongraphical text editor in the world. Those that aren't built in are provided by libraries of programs written in a LISP-like programming language. There is essentially *nothing* that Emacs can't be made to do, with enough clever programming.

For the novice, however, Emacs presents a fairly imposing barrier. All of the basic text editing commands are invoked by simple but unintuitive keystrokes. Although hundreds of thousands of people may by now think it is "natural" to hold down the control key and press the v key simultaneously to see the next page of text, this is not really a basic human instinct, and each new Emacs user must learn it for him- or herself. Worse, the remaining keystrokes are bound to nonessential features, some of them so arcane that it can take hours to explain to new users what they are *for*, much less how to use them. When a new user accidentally types the control-x–control-u key sequence, for example—and this isn't as unlikely as it may sound to those unfamiliar with Emacs, because many important commands start with control-x—this invokes a general and sophisticated "undo" mechanism, which will try to undo the most recent actions of the user. If the user doesn't understand this fact, however, it will appear that his or her text has been modified in a frivolous and unpredictable manner, and it will, in fact, have a radically different effect each time the user types those keystrokes.

(Worse still, if he or she hits the space bar afterward, this is interpreted to mean "undo even more.") For the novice, Emacs is far too complicated, filled with features that are not only inessential, but deeply confusing as well.

This is, in fact, a glaring deficiency of many powerful programs—Emacs is far from unique in this regard. Complexity and power go hand in hand. To many programmers, this seems like just a fact of life that users will have to learn to deal with. They tell users all too frequently that they have to choose between programs that are powerful and programs that are simple to use. This is by now so widely believed that there are essentially separate markets for "simple" and "powerful" computer programs.

The idea that a powerful program has to *appear* complicated is, however, utter nonsense. The appearance of a program is almost entirely separable from its internal complexity. What *is* true, however, is that the *easiest* way to write a powerful program is to make it appear complicated. In fact, making a powerful program appear simple usually makes it even more complicated on the inside, but, if done correctly, hides this complexity from the users.

An electric lamp, for example, appears to most people to be an extremely simple device. By turning a switch, you can turn the light on or off. If this doesn't work, you can check to see that the lamp is plugged in, change the light bulb, and check the circuit by plugging something else into the same socket. This is all that most people ever learn about electric lamps. On the rare occasion that none of this knowledge suffices to turn a light on, they either declare the lamp broken and throw it away, or (in the case of a lamp mounted on the wall or ceiling) call an electrician to fix or replace it.

Those with a bit more understanding, however, have a few more options. By taking the lamp apart they may find loose wiring and repair it. By bypassing the switch they may be able to determine that the problem is in a defective switch. Like the electrician, they may repair or replace defective wall fixtures as well.

Those with still more understanding may not, in this case, have any more options, but they will nonetheless have a deeper appreciation for the whole process. They may be knowledgeable enough about power generation and transmission to explain the nuclear reaction at a power plant that is almost simultaneouly generating the power that feeds the bulb. They may be able to explain the physics of how the filament in the bulb glows to produce light. In the simple case of a defective light bulb, they may be able to account for the defect in terms of their knowledge of the intricacies of light bulb manufacture.

The point is that a "simple" light bulb hides within itself an incredibly complex base of technology, knowledge, and expertise. Even more dramatic is an automobile, which has controls so simple that nearly anyone can drive, but ever-increasing complexity and sophistication concealed beneath the hood.   The beauty of this complexity is that it is completely hidden from those who don't need it, but remains available for those who do—for example, for the tinkerers who like to fix their own cars.

Looking Simple Is Beautiful

There are, it turns out, quite a few examples of programs that manage to be extremely complex without seeming so to the users. A prime example of this is the document production software known as Scribe.[1]

Scribe is an extremely sophisticated system for producing formatted text documents on a wide variety of printing devices. It includes a staggering richness of features, including programming mechanisms for defining whole new types of formatting.   It is, however, primarily oriented toward nonexpert users. The user can say, at the beginning of a document, "@make(thesis)," to tell Scribe that this document should be considered a thesis. This invokes a wealth of knowledge about what theses look like, and produces a document with appropriate margins, page numbering, sectioning, and so on. The expert, however, can make minor or major modifications to the definition of "thesis" to produce precisely the desired result.

This programmable richness is precisely what has allowed Scribe to come to do the right thing so often, for casual users, by default. The current defaults are the product of years of evolution, in which sophisticated users continually modified the default settings in search of a good description of what documents should look like. But this programmability does not come for free; a high cost is paid in complexity of the software. Indeed, the Scribe code is far larger than

---

[1]Scribe is a trademark of Scribe Systems, Inc.

most comparable document production systems, even though the others may seem much more complex to those who are trying to use them.

The real secret of Scribe's success is that it has successfully *hidden* its complexity from most of its users. To them, it seems simple. Indeed, after the second major release of Scribe, all of its programming features were actually removed from the Scribe user manual, and were relegated to a secondary manual that most users have never even seen. Thus the users don't even have a huge instruction manual to contend with, so completely has the complexity been hidden from them. Most Scribe users haven't any inkling of how complicated Scribe really is, and this is just what the implementors wanted.

A more familiar but less extreme example of this phenomenon is the Macintosh operating system. Casual users will uniformly describe the major competitors of the Macintosh (MS-DOS, for example) as being much more complicated than the Macintosh, when in fact the reverse is true when one views the internal operating system code. In order to achieve its much-heralded simplicity of interface, the Macintosh required a more complex and less maintainable basic operating system, but it effectively hides that fact from its users in almost all cases.

Complexity is inherent in user interfaces. In the old world of teletypes, getting a file name from a user was a simple matter of reading a line of text input. In the graphical world of the Macintosh, for example, getting a file name from the user may involve a dialog box, a mouse-selectable hierarchical list of files, and so on. It should surprise no one if such "friendlier" interfaces are inherently more complex.

There is a slogan, in software engineering circles, that is so widely accepted that hardly anyone ever bothers to state it any longer; it is referred to simply as the "KISS" rule. KISS stands for "Keep It Simple, Stupid." It refers primarily to the tendency of programmers to be overly clever, to destroy their programs by writing code so ingenious that it can never be fully understood or debugged. Unfortunately, applying KISS to user interfaces often produces interfaces that are simple *and* stupid. In user interfaces, it is more often best to provide power for the users who really need it, but to hide, as completely as possible, the resulting complexity from new or casual users. To the user-interface programmer, KISS is an incomplete philosophy and a misleading slogan.

One appealing consequence of hiding complexity is that you don't ever have to win an argument about the basic philosophy. The casual users (often, as mentioned above, powerful managers) who decry "overly complex" programs don't need to be convinced that they are wrong—that is, that complex programs with simple interfaces are the real solution. If they are presented with good examples of such interfaces, they will naturally assume that their advice was followed. (Afficionados of Scribe often say they like it "because it is so simple.") The program they are using will appear to be a simple program, and nothing more will need to be said. But if ever their use of the program grows more demanding, the power will be there waiting for them.

Of course, the KISS philosophy still makes a great deal of sense from a software engineering perspective. Complex programs are undeniably harder to maintain, and indeed our example, Scribe, is notoriously bedeviled in this regard. Clearly complexity should be avoided wherever possible. However, people are complicated, and making computers deal with people is inherently complicated. In the case of user interfaces, this must always be balanced against the lure of simplicity.

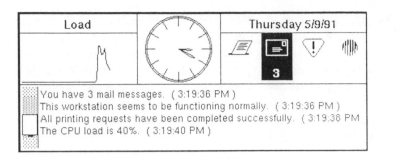

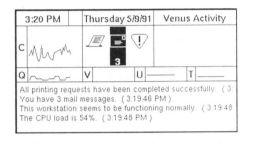

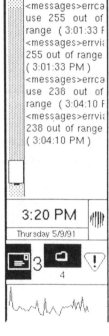

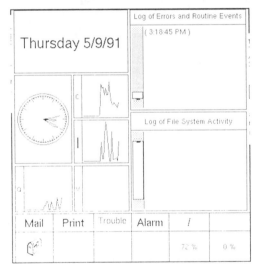

Some sample screens of the Andrew Console program. Console is an example of a highly configurable program that can be customized to create an infinite set of status-monitoring windows. Despite this flexibility, a substantial amount of effort was still required to design the default console display, shown at the top, which is generally comprehensible even to very naive new users.

Chapter 10

Tune Defaults to the Novice

I had rather have a fool to make me merry, than experience
to make me sad.

—William Shakespeare

The Primal Scream

Very few people who write computer programs for a living are at all
frightened of computers. Such people tend to be true afficionados,
even connoisseurs, of computer software. It is inevitable that they
view the human encounter with software differently from those for
whom interaction with a computer is at best a useful time-saver, and
at worst a fearsome adventure in ignorance and humiliation.

Sitting down to use a new computer program is, for most
nontechnical people, intimidating. Particularly if the person is not
merely using a new program, but a whole new *type* of program, the
experience can be deeply traumatic, rather like taking a standardized
test for admission to college. Even highly sophisticated users often
experience anxiety when confronted with a new program, and will
cling to their old favorites with impressive determination.

Thus the first hurdle that all software must face is simply getting
people to try it out for long enough to overcome this initial period of
anxiety. But even assuming that this hurdle has been passed, another
quickly becomes just as important: the software and documentation
must avoid overwhelming users as they learn the system. Like a

wildlife biologist observing the mating rituals of reticent birds, the user-interface programmer must quietly stalk users, exercising the greatest care not to frighten them away during the initial approach.

Often, a new program will confront the user with an array of specialized terminology, either entirely new or containing old words with new specialized, technical meanings. Similarly, a graphical program may, through sheer visual complexity, give the user the feeling of having been seated at the controls of a fighter jet and told, "Go ahead; take her out for a spin." The initial impression can be overwhelmingly frightening—a fact technical experts often forget.

A good rule of thumb is that if users can't figure out how to do *something* that seems useful in the first thirty seconds, they may never want to touch the program again. This is a principle successfully adhered to, over many decades, in the design of the controls of automobiles. Each successive generation of automobiles has introduced changes in such controls, but the basic design has evolved gradually, so that users of new cars do not experience the same kind of "stage fright" that is so familiar to the users of new software. If a 1950s Rip van Winkle sat down in a 1989 model car, he might have a bit of trouble getting the windshield wipers or lights to turn on properly, but he wouldn't be panic-stricken or unable to drive the car. Notably, he might not have the foggiest notion about how to use the car's most interesting innovations, such as the compact disc player, but this wouldn't prevent him from using and enjoying the car.

Of course, software designers have a much harder task: their audience is fairly unlikely to be familiar with a similar product, although this applies with decreasing frequency to such standard software as text editors. Moreover, the inherent plasticity of the software medium suggests endless variations on the standard themes. Such variations are wonderful in principle; they are the basic engine of software evolution. But they are misapplied when they interfere with that first fear-laden encounter between humans and software.

The easy solution, of course, is to make software so simpleminded that any fool can use it. This certainly solves the problem, but at the cost described in chapter 8—the inevitable recognition, by the increasingly sophisticated user, of just how simpleminded the program really is. This is, ultimately, not a solution at all. It insults and frustrates the user, and it leaves no outlet for much of the programmer's creative energy.

Don't Worry about That Just Yet

The *correct* solution, on the other hand, is more complicated. Programs, however complex, must be designed to *look* simple at first glance, even if this means that substantial portions of their functionality are effectively hidden from the new user. Ideally, the complexity of a program's interface and capabilities should match itself automatically to users' needs.   As users grow more sophisticated, they should be presented with an increasing range of functionality. This approach has been described in some detail in the literature, and is often referred to as the "training wheels" approach (Carroll and Carrithers 1984).   It can be found in a number of commercial products, at least in the form of simple and extended sets of menu items.

This solution simplifies the way novices learn to use software, but at a substantial price—it typically entails a whole new layer of software to manage the different "levels" of interface. In practice, it is essentially impossible, with current technology, to match the interface level to the user's expertise level completely automatically. Instead, most programs that attempt to offer such a multileveled interface require some user-initiated action to turn on the more sophisticated features. Such actions are most often effected through an "option-setting" mechanism, which may itself vary enormously in the power, simplicity, and sophistication of its interface.

Of course, it is inappropriate to confront new users, just daring for the first time to try to change an "option" setting, with an overwhelmingly complex interface for setting options. Thus, the problem of interface layering is recursive—if an option-setting interface is sufficiently complex, it should itself be made available in simplified form for novices, with one of the options presented being the option to use the more complex and powerful option-setting interface. Although few programs go this far, many of them have such complex customization interfaces that they should consider such layered customization.

Although it takes a somewhat different form, this same strategy has been used with great success in other user interfaces—notably, consumer electronics. Many video recorders now present a rather simple interface to casual users, with only a few clearly labeled buttons to press. Only after they get home and try to figure out how to do something more complicated will consumers typically find the extra little door that conceals a dozen or so more cryptically labeled buttons. Eventually, one has to figure out what most of them mean, but hiding them until the basic functions are understood dramatically reduces the fear factor—at least in the store showroom.

How Do They Know What to Learn?

If a program looks simple to new users, how are they to know what options exist, and what hidden capabilities the program might have? All too often, features hidden behind "options" fall into disuse through obscurity, a fate to be regretted. Somehow, then, a program's simple interface must include some kind of "hint" as to the program's full functionality.

Ideally, a program could solve this problem through the use of user modeling and other artificial intelligence techniques. Such a program might remark, after watching a user struggle with the novice interface for a week or so, "By the way, if you set the XYZ option, you can do this kind of thing much more easily." But although such advice-giving systems have been demonstrated in research labs (Finin 1983), their computational properties are such that they may not ever be practical for production-quality user interfaces.

This is not, it turns out, such a great loss; such advice can be unnerving, and is often wrong anyway. A much simpler, but still effective, technique is simply to include an option-setting command prominently as part of the novice interface. This is particularly easy in the case of a menu-based interface, where a simple "set options" menu may be all that is necessary. When users select such an option, they can find themselves in something as simple as a question-answer dialog about features that can be turned on, or something as complex as an online tutorial in the use of a programming language to extend the functionality of the system. The novice interface really only needs to include a little "hook" to attract the interested novice to explore the deeper complexity of the system.

The Andrew Console Program

A good example of the "layered" approach can be found in the Andrew Console program. Console is a program originally designed to monitor a few vital statistics about a user's computing environment, and to display them in a pleasing fashion on the screen. For example, Console can tell the user the time, the current load on the machine, whether the user has mail, how full the disks are, and so on. It was designed as a replacement for a whole suite of smaller programs, each of which monitored one or a few of these items.

It was realized early on that different users would have widely varying uses for such a program. Some would want to monitor all sorts of arcane system data, while others would want only the highest-level information. Some would want the time displayed with a digital clock, while others would prefer analog. The possibilities grew quickly, and it seemed natural to make Console highly configurable.

This was done by making Console *programmable* in a simple programming language. Using that language, one could design a wide range of visual interfaces for the Console program, a few of which are pictured at the beginning of this chapter.

However, for most people, the phrase "a simple programming language" is an oxymoron. Clearly a version of Console that *required* its users to program it would be inaccessible to the majority of its intended users. Indeed, in its first release, Console was rejected by many potential users because they didn't like the default display and couldn't be bothered with designing their own.

For this reason, a great deal of effort was ultimately put into developing a *default* Console display that would be intuitively clear and useful to novices. When users ran the Console program without any programming or customization, they saw the standard screen image. (Though most Console users eventually switched to something other than the default screen layout, the default lured them into using Console in the first place.)

Unfortunately, the programmability set up something very much like a brick wall between the sophisticated and unsophisticated use of the Console program. If you didn't want to write a program, you got the default version. If you didn't like the default version, you had to learn a whole new programming language—a rather big plunge for the casual Console user to take.

An intermediate alternative was provided by the creation of a *library* of Console programs. Simple menu items allowed the user to switch back and forth among the various Console displays in the system library, in search of the "ideal" Console. Once that ideal (or a reasonable approximation) was found, a simple option could be set to declare it the user's preferred default Console.

Crucial here are the multiple layers of customization behind the novice-oriented defaults. By default, users received the simplest, most easily understood display that the designers could devise. Using simple menu options, users could select from among a relatively small number of predefined alternative displays in the library. Using a programming language, users could design a whole new display. Given these multiple layers through which users could progress to become gradually more sophisticated, a surprising number of people actually ended up learning the language well enough to write their own Console files. It is doubtful that so many would have learned the language had they been unable to experiment first with the simpler customization mechanism, and it is even more doubtful that they would have gotten far enough to do that experimentation had the initial display not been so intuitive and comprehensible.

Who Knows Novices?

Unfortunately, a determination to tune defaults to the novice, though crucial, is not enough to guarantee success in that effort. Almost inevitably, nobody on the software design will really match the profile of a novice user, and in any event not all novice users are alike. How, then, can the developers be sure that their idea of novice defaults have some connection to the reality of novice users?

As in many other aspects of user-interface development, the answer is simple to state and tedious to execute: there is simply no substitute for trying things out on real users in the target group. However the novice-oriented defaults are initially designed, they are probably not going to work for novices the first time. Only by trying them out on real novices, and then making changes based on the problems those novices encounter, will the defaults gradually reach a state where they are really "tuned" to the novice user.

In the case of the Console program, the developers went so far as to set up a committee, on which developers, documenters, and user consultants worked together to determine the right default settings.

Even then, it was necessary to iterate through a series of designs before a default Console was designed that really met the needs of first-time users.

Software should be tuned to novices so that they can learn to use it quickly, but the developers are eminently unqualified to understand what novices will need. This is a fundamental reality of software development—and though nothing will "cure" this problem, an understanding of this reality will increase the likelihood that a process can be found for arriving at the right default settings.

```
┌─────────────────────────────────────────────────────────────────┐
│ ⊠  ez /fs/pen/usr2/nsb/tmpsm/sendmsg.c*  ▓▓▓▓▓▓▓▓▓▓▓▓▓▓▓▓▓▓▓▓▓ 囘 │
├─────────────────────────────────────────────────────────────────┤
│ #define SM_STATE_SENDING 4                                        │
│ #define SM_STATE_SENT 5                                           │
│ #define SM_STATE_VALIDATING 6                                     │
│ #define SM_STATE_VALIDATED 7                                      │
│ #define SM_STATE_VALIDATEFAILED 8                                 │
│                                                                   │
│ /* constants for the Deliver() subroutine */                     │
│ #define FORCE_ASK_ABOUT_FORMATTING 0                              │
│ #define FORCE_SEND_FORMATTED 1                                    │
│ #define FORCE_SEND_UNFORMATTED 2                                  │
│                                                                   │
│ extern char *index();                                             │
│                                                                   │
│ #ifndef _IBMR2                                                    │
│ extern char *malloc();                                            │
│ #endif /* _IBMR2 */                                               │
│                                                                   │
│ void sendmessage__LinkTree(self, parent)                         │
│ struct sendmessage *self;                                         │
│ struct view *parent;                                              │
│ {                                                                 │
│     super_LinkTree(self,parent);                                  │
│     if (self->SendLpair != NULL) {                                │
│         lpair_LinkTree(self->SendLpair, self);                    │
│     }                                                             │
│ }                                                                 │
│                                                                   │
│ void sendmessage__UnlinkTree(self)                               │
│ struct sendmessage *self;                                         │
│ {                                                                 │
│     super_UnlinkTree(self);                                       │
│     if (self->SendLpair) {                                        │
│         lpair_UnlinkTree(self->SendLpair);                        │
│     }                                                             │
│ }                                                                 │
│                                                                   │
└─────────────────────────────────────────────────────────────────┘
```

EZ, The Andrew multimedia editor, editing a program in the C language. Note that the closing brace, which the user has just typed, has caused EZ automatically to highlight the code fragment that was just completed. The EZ features oriented toward C programming are an example of catering to the needs of specialized but important experts without compromising the integrity of a novice-oriented program.

Chapter 11

Don't Neglect the Experts

Is there anyone so wise as to learn by the experience of others?

—Voltaire

The shoemaker makes a good shoe because he makes nothing else.

—Ralph Waldo Emerson

We have seen that user-interface software must be tailored to have simple defaults for novices, but should have sufficient power and flexibility to meet the needs of users as their expertise and sophistication grows. Left unmentioned so far are the needs of real experts, the small group of hard-core users who will squeeze every drop of performance and functionality out of a system and clamor for more.

Experts have, in recent years, gotten a bad reputation among those who worry about user interfaces. In fact, one of the worst insults one can make about a user interface is, "This is obviously designed for experts." With the rise of the Macintosh, and of "friendly" software in general, it has become a truism that "good" software is software that can be used by nonexperts. The unfortunate (and erroneous) conclusion that many have drawn from this is that a consideration of experts has no place in user-interface design.

What, though, is an expert? An expert is a person who used to be a novice, but has learned something—more, in fact, than most people have learned. In many situations, an expert is precisely what you hope a novice will become after using your software for a while. Why, then, would it make sense to neglect the experts entirely?

Fallacies About Expert Users

There are at least three common reasons why interface designers avoid worrying about experts, all of them spurious.

If You Do It, You'll Go Blind

The basic contention here is that thinking about experts warps your mind. This is essentially a reaction against all the horrendous interfaces that have been usable *only* by experts. Nearly anyone who has worked with computers has encountered dozens of such programs, and has loathed them. It is therefore understandable to fear that designing with experts in mind will lead to more such programs.

The fallacy, however, is that those "expert-only" programs were typically *not* designed as user interfaces for experts; they were simply not designed as good user interfaces at all! Such programs are, typically, built for functionality rather than for interface, with the result that the user interface evolves incoherently and without any concern for ease of use. That they end up being considered "expert-only" programs does not reflect the intent with which they were designed. Rather, this is more a consequence of a complete lack of design, and it is their very lack of a well-designed interface that makes them unusable by most mortals.

Designing a program with experts in mind does not, of course, guarantee that nonexperts will be unable to use it. However, designing a program without *any* users in mind is almost certain to guarantee that *at most* a few experts will really be able to use it. Such "undesigned" systems are probably the source of most of the "expert-only" interfaces in the world.

The Experts Can Take Care of Themselves

The idea here is that experts are, by definition, sophisticated users who can cope with nearly anything. After all, experts have learned how to use some of the most obscure interfaces in the world. Therefore, it seems, the interface designer can safely ignore expert users, confident that they will be able to muddle through with whatever interface is finally built.

This reasoning is true so far as it goes, but it leaves out a crucial fact: experts tend to have (and be aware of) more alternatives than novices. Moreover, experts are more likely to be established users of other, possibly competitive systems. The problem with experts is not one of *capability*—of course they will be able to figure out how to use your novice-oriented systems, if they're so inclined. The problem with experts is one of *salesmanship*: as the program designer, you need to persuade the experts that it is in their interest to use your program.

In fact, experts can take care of themselves so well that they will often conclude prematurely that they don't need your software at all. Since experts commonly occupy positions of authority, this conclusion can have repercussions for nonexpert customers as well. If expert users conclude that your software is not worth the trouble, they may be in a position to prevent or discourage some of the novices who constitute your "target audience" from using your software, as well.

They're Not Worth the Money

Experts are relatively few. The big audience is the great unwashed mass of nonexpert computer users. Since the ultimate goal of the software developer is to sell a product, thinking about experts is a waste of time.

The relative truth of this claim is less clear than the others because cost-benefit analysis is such a slippery endeavor. Certainly it is true, for example, of educational software aimed at the elementary school market, as there are few true experts to be found in primary grade classrooms. On the other hand, the claim is ridiculous when applied to air traffic control or airplane cockpit software. The software's audience is crucial here.

Still, the importance of experts, even from a marketing

standpoint, seems to be widely underestimated. As mentioned previously, experts often have a disproportionately large role in decisions about software purchasing. Even novice-oriented word-processing software would do well to have a few eye-catching features to get the attention of the customer's local "software expert," who will likely be consulted when it comes time to decide what word-processing system to buy.

Give Shortcuts to Experts

How much attention, then, should be paid to the expert user in the process of designing user interfaces? As usual, there are no easy formulas to answer the question. The key variables are the nature and relevance of expertise in the task domain.

In general, expertise in the use of software consists of knowledge of various kinds of shortcuts. Some shortcuts are essentially physical, like the knowledge that typing "control-x–control-c" accomplishes the same thing as selecting the "quit" menu and confirming it by typing "yes" and pressing the enter key. These shortcuts save expert users time by minimizing the physical actions that they have to take to get their work done (e.g., by minimizing the number of keystrokes). The price they pay (their expertise, if you will) is the requirement that they memorize more specialized and typically nonmnemonic commands.

More interesting shortcuts are often conceptual rather than physical. To use these shortcuts, the expert needs not merely to memorize additional commands, but to assimilate the concepts that permit the use of the advanced commands. For example, in text editors, the notion of a macro definition—using a series of basic commands to define a single complex compound command—is an extremely powerful shortcut, but one that cannot be learned in terms of simple memorization.

The question of designing for experts, then, is primarily the question of what shortcuts should be provided. But there is one additional aspect worth considering as well—the question of advertising. In addition to providing experts with shortcuts, you must find a way to let the experts know that the shortcuts are there without cluttering up and confusing the basic interface for novices.

A good example of how experts can be kept happy without

destroying a system for novices can be found in another tale from the history of text editors in the Andrew Project. In chapter 8, we saw how the first Andrew text editor, EditText, failed largely due to lack of power. This left a situation in which the user community was highly skeptical about new text editors on Andrew, and seemed more entrenched than ever in its use of the venerable Emacs text editor.

Thus, when the new editor, known as EZ, was introduced, it was greeted with something less than wild enthusiasm. It had most of the same fancy features as EditText, but with an extensible internal structure designed to permit it ultimately to become even more flexible and versatile than Emacs. The problem, of course, was in convincing the users that the pursuit of this promise was worth the sacrifice of abandoning their old friend, Emacs.

A few users, of course, were easily persuaded. In particular, the few loyal EditText users were relatively easily converted to EZ, which was an obvious step forward. The Emacs users were tougher to persuade, and toughest among these were the most hard-core Emacs users—the programmers themselves.

The situation had, to put it frankly, a strong potential for embarrassment for the EZ developers. EZ had been developed over several years, and at substantial cost, as "the Andrew editor" and as an embodiment of the next generation of editing software. There were strong technical reasons to prefer EZ to Emacs. Yet prominent among those who persisted in using Emacs instead of EZ were most of the programmers for the Andrew Project itself! (The EZ developers generally used EZ, but there were many other parts of Andrew besides EZ, and the other developers were still using Emacs.) If EZ was to succeed, that expert audience had to be one of the first to be converted, even if it constituted a relatively small portion of the eventual target audience of the software—if only because their refusal to use it would be such a public embarrassment.

The EZ developers could have pushed for a political solution. They could easily have required that all EZ developers refrain from using Emacs, and with a political struggle they might even have been able to force the rest of the Andrew programmers to use EZ as well. But this Stalinist solution would have done little to persuade the rest of the world to use EZ. Instead, and very wisely, the EZ developers tried to analyze the roots of the programmers' reluctance to switch to EZ.

Dozens of reasons were offered by the developers themselves.

Disturbingly, it seemed that each developer had his or her own list of reasons, with little overlap among them. Emacs had been around for over a decade, and had accumulated an enormous number of features, great and small, that the experts had come to love and cherish. Though no single feature would have been hard to add to EZ, duplicating all of them was out of the question in the short term.

Instead, the EZ developers decided to balance the scales. Against the various minor features that Emacs offered and EZ lacked, EZ would offer a few key features that could not possibly be provided by Emacs. This would serve a dual purpose: it would lure the programmers to EZ, and it would demonstrate clearly the superiority, the ''next-generation'' nature of EZ. The programmers who began to use EZ would grumble about the missing features they had grown used to in Emacs, but they would recognize that those features could be added in time. Indeed, some of them might actually take the time to add the features themselves—such is the value of having the experts on your side.

The feature that was selected for emphasis was an ''editing-mode'' package for C programs. A popular Emacs package, known as ''electric C,'' was commonly used by programmers to simplify the writing and editing of programs in the C programming language. This package helped the programmer to check for balanced parentheses, to indent the program nicely to make it more readable, and so on. The EZ developers built a package known as ''ctext'' that did all these things and more. In particular, it automatically analyzed the syntax of the C program and displayed appropriate parts of the program in specialized fonts. Comments were displayed in italics, and the names of procedures being defined were displayed in bold. The graphical display of the program's structure was a great boon to the programmer. For example, with ctext it was almost impossible to leave some of your program commented out by accident, as the italics would be so noticeable on the screen.

The appeal of ctext to programmers was such that, instead of grumbling about EZ's inadequacies and continuing to use Emacs, they grumbled about EZ's inadequacies while actually using EZ. This led to a quickening of the pace of improvement in the EZ program itself, and thus ultimately to the success of EZ with a wider and more diverse community. Although EZ was never a program designed primarily for experts or programmers to use, the existence of a few key features designed for those audiences was nonetheless a crucial factor in the program's success. Moreover, it is worth noting that since these features are only invoked when C programs are edited, they imposed absolutely no penalty on naive nonprogramming users. Sometimes, one really can have one's cake and eat it, too.

```
┌────────────────────────────────────────────────────────────────────────────┐
│  andrew   (No such message directory)                                        │
│ ┌─                                                                           │
│ │                                                                            │
│ │                                                                            │
│ │                                                                            │
│ │                                                                            │
│ │                                                                            │
│ │          ┌──────────────────────────────────────────────────────┐         │
│ ┼─         │   andrew no longer exists. Delete your subscription?  │         │
│─┤  Click   │                   ┌─────┐ ┌────┐                      │         │
│ │          │                   │ Yes │ │ No │                      │         │
│ │          └──────────────────────────────────────────────────────┘         │
│ │  Click the right mouse button on a message directory name to see the titles of all messages in │
│ │  that directory.                                                           │
│ │                                                                            │
│ │  Click the left mouse button on a message directory name to see the titles of the new          │
│ │  messages in that directory.                                               │
│ │                                                                            │
│ │                                                                            │
│ ┼─                                                                           │
│ └─                                                                           │
│ Checking for new messages on andrew. (3.18.36)                               │
└────────────────────────────────────────────────────────────────────────────┘
```

In the Andrew Message System, if a bulletin board ceases to exist, each subscribed user is informed of this event by a dialog box like the one shown above. Unfortunately, the mechanism by which the system concluded that a bulletin board had ceased to exist was occasionally buggy in the early versions of the system. In this case, the bug hit on an unfortunately named bulletin board—the one called simply "andrew." The resulting dialog box was certainly startling, to say the least, to all the regular Andrew users who saw it.

Chapter 12

Your Program Stinks, and So Do You

"Your program stinks, and so do you!"
—another satisfied Andrew user, in a private mail message to the author. The user was responding to being told that a new feature he had requested would be prohibitively difficult to provide within the context of the current implementation.

Fools and young men prate about everything being possible for a man.

—Soren Kierkegaard

Nearly anyone who has supplied a user-interface program to a significant number of users has been on the receiving end of a comment similar to the title of this chapter. User interface software is in this regard very much like art—no matter how good it is, there will always be someone who despises it. This a simple and unpleasant fact of life for user-interface designers, most of whom gradually learn to be somewhat thick-skinned as a result.

The phenomenon is entirely natural and understandable. First of all, user interfaces inherently involve questions of style and taste. What one user sees as an aesthetically pleasing display is seen by another as an outrageous waste of screen space. What one user sees as a clear, comprehensible error message is, to another, patronizing in tone and insufficiently informative. Designing a user interface to please everyone is about as likely as composing a symphony that everyone will enjoy.

Second, and more important, the user interface is virtually the only aspect of a computer system that most people are brave enough to judge. The average user knows that he can't really offer an intelligent opinion about a computer's architecture, or about the design of its network protocols or disk controller. But when it comes to user interface, the average user can say, "I may not know programming, but I know what I like." The mere ability to criticize a program without fear of looking utterly stupid is *empowering* for users—it makes them feel more in control of their situation. This is one reason why some users (the ones who seem to feel the most need for exercising power over their computers) complain endlessly about the interfaces of the programs they use. In a sense, the interface designer may be doing such users a favor, psychologically speaking, by giving them something they feel they can safely complain about.

Yet the most important reason why user interfaces seem to face an endless stream of complaints and suggestions for improvement is even simpler and more basic: there is usually an endless stream of inadequacies in the interface, and it is always easy to see genuine room for improvement. In other words, when the users tell you your program stinks, they're usually right! By the time a program is built and people have a little experience using it, it is nearly always obvious how the program could be improved. That the improvement may be difficult (because it involves radical redesign of the software) or may not have been so obvious before the program was built is irrelevant to the user, who can only see the "obvious" flaws, not their historical or technical justifications.

In short, your program really *does* stink, and the sooner you get used to the idea, the better. The inadequacies of your software are simply a reflection of your frail, shortsighted, and limited human nature. Every program ever built is doomed to eventual obsolescence. In the case of user interfaces, however, this is somewhat more painful because the failings and obsolescence are on public display. The seeds of eventual obsolescence are present from the very first release of the software, and each complaint or even friendly suggestion from a user can feel, to the sensitive developer, like the twisting of a knife, or like a hammer, nailing one's own coffin permanently and prematurely shut.

Clearly, a good attitude is extremely important. Maintaining such an attitude is not an easy thing, however. Inasmuch as user-interface designers are artists, they share with all other artists the fundamental existential burden of knowing they are creating a finite, temporal work that will last a short time and then vanish as tracelessly as yesterday's sunset. The unique dilemma of artists-as-interface-designers is that their audience—the very people for whom they create their art—seems to work feverishly to hasten its demise, through the rash of suggestions and complaints that lead inevitably to the next generation of the software. It is as though a crowd of people gathered around Michelangelo, urging him ceaselessly to make changes, small and large, to his sculpture, until at last he knew that he had chipped away too much to be able to continue with this particular work, and so had to start all over.

The real problem, then, is one of mental health. It is very easy for the interface designer to respond to the suggestions with an orgy of improvements, in the ultimately hopeless attempt to "satisfy" the users once and for all. But the users will not, in the end, be satisfied. Software is endlessly improvable, and as it evolves it inevitably reaches a point where a rewrite—either by the original author or by someone else—is the only way to make the next round of improvements. Ultimately, the orgy of improvements only hastens this process, by accelerating the software's evolution to the point at which a new program is required. In this sense it is not a bad thing—by hastening the evolution, it gives users better software sooner—but it can be very hard on the poor programmers. Programmers frequently "burn out" under such pressures, and a few bad experiences of this kind are a major reason why so many good programmers avoid or simply refuse assignments involving user interfaces.

As with many crises of mental health, the solution may lie in a better understanding of the phenomena that are perceived as "the problem." In this case, the real problem is not that programs are never good enough, but rather that programmers approach them with at least a subconscious desire to make them perfect and eternal. Such goals are the stuff of which great art is made, but are also, as even the most successful artist knows, doomed to failure. Coming from relatively narrow technical backgrounds, many programmers don't even realize that what they're doing when they create a user interface is essentially artistic in nature. It is not surprising, then, that they are unprepared when they stumble into the anguish that is fundamental to art itself.

Why, then, don't artists burn out the way programmers do? The short answer, of course, is that they all too often do. The history of art is littered with great artists whose careers ended at a very young age, often even in suicide. But most artists manage to avoid this fate, in large part through the warning example of their less fortunate colleagues. An awareness of the fundamental futility of the endeavor is one of the keys to the artist's sanity, as artists have recognized throughout the ages. This awareness may be reached through religion, as in the recognition of human insignificance in the face of the Infinite, or through a more fatalistic attitude. Almost any approach, it seems, will do. The danger is greatest only when it is unrecognized, when a programmer is consumed with the impossible passion to "perfect" a program without any deeply felt awareness or understanding of the futility of this passion.

Perhaps in a few years or more these reflections will seem silly and unnecessary, as it will be generally recognized that programming, as a human activity, shares this essential futility with most other arts and crafts. But today, in the early years of the computer age, programmers often embark on their careers with little or no exposure to the philosophies and attitudes that enable people to proceed under these circumstances. To hold up under the strain of the inevitable death-chorus of hungry users, a user-interface programmer cannot long survive without a dose of philosophy that is rarely to be found in the engineering curriculum. (Of course, programmers are spared this anguish when they are so isolated from their users that they never hear any complaints, but this solution is worse than the problem, implying as it does that user feedback has little or no role in the software's evolution.)

A good attitude to take, from the first day of any programming project, is that the system being built is fundamentally flawed and doomed. The goal of such a project, then, is simply to build a system that will last long enough for a better one to come along, and perhaps also to be, for a brief moment suspended between eternities, the best program of its kind yet built.

When viewed from this perspective, the inevitable demise and abandonment of the software is a good thing, because it means that it has done its job and something better has come along. Often, one can arrange things so that the replacing software is also one's own; there is a peculiar satisfaction in driving the nail into one's own coffin, and it is surely less painful for a programmer to see his software abandoned if he played an active role in creating the system that replaces it.

With such an attitude, programming is a constant process of new beginnings. As the chorus of suggestions from users arrives, one pushes hard to incorporate as many of the good ideas as possible. When, inevitably, this grows harder and harder, the correct response is to step back and consider how the system could be redesigned from scratch to incorporate the new ideas. Eventually, as enough of these ideas accumulate, it will simply be more appealing to start over than to work on the old system. This is the point at which the *designer* is done with the system. After this point, *maintainers* may keep the software running for years, if it is important enough, but the designer will have long since moved on to a new effort, doing his or her best to learn from past mistakes. This is why designers will often have virtually nothing good to say about the systems they have built in the past.

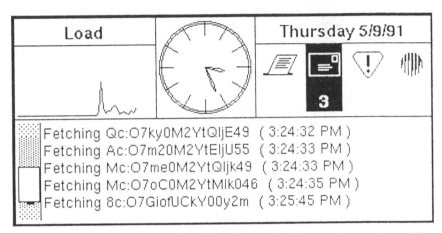

In the above picture, an algorithm for producing random unique file identifiers has produced an obscenity, leaving the impression of a program that is swearing at its users.

David Romig reported a similar incident regarding unique identifiers used in mailing labels for the alumni records office of a large university. The word "fuck" appeared on one of the mailing labels, causing severe distress in the alumni office and requiring new action on the part of the programmers.

"Thereafter we were instructed to add the 'dirty-word-routine' which performed a table lookup of every word that a committee of about a dozen of the raunchiest people in the department could come up with. But what about short phrases? And how about maintenance of the table? Whose budget does this come out of? A student programmer, invited to a meeting to 'see design in the real world,' made an unwanted suggestion: 'Just convert to base-31 and don't use vowels.' It worked." Without vowels, dirty words could not be generated, and it was no longer necessary to form an exhaustive list of all the dirty words that might be checked for.

Chapter 13

Listen to Your Users, but
Ignore What They Say

Truth comes as conqueror only to those who have lost the art of receiving it as friend.

—Rabindranath Tagore

You must act in your friend's interest whether it pleases him or not; the object of love is to serve, not to win.

—Woodrow Wilson

As previously mentioned, there is never a shortage of suggestions and complaints from the people who use software. These complaints often come from people who are quite sure they know not only what the problem is, but how it should be fixed as well. To hear the complaints, one could easily believe that improving a user interface is a simple matter of getting some moderately intelligent person sufficiently involved to tell the stupid programmer what things obviously need to be done.

Unfortunately, even some programmers seem to believe this, and will happily follow even the most senseless of instructions. The world is full, accordingly, of bad user interfaces that were essentially designed (or redesigned) by nontechnical people with no real idea of what they were doing, and implemented uncritically by programmers who were, like Adolf Eichmann scheduling deportations to the Nazi death camps, "just following orders." To take directions uncritically from the user community (or even from management, as will be discussed later) is, for the interface designer, a shirking of professional responsibility.

A good architect, if urged to make the exterior of a new office building bright pink with purple polka dots and a giant happy face in the middle, will resist these urgings, regardless of their source, out of professional pride and a determination to produce a good-looking building *despite* the expressed wishes of his patrons. This is not, it must be emphasized, a simple matter of willful disobedience to those who pay the bills. The architect knows that it is likely that everyone, including the people who now think they want the happy face, will quickly get sick of them. He exercises his professional judgment on behalf of the client, even when that judgment conflicts directly with the client's perceptions of his own needs.

Such is often the situation with user-interface software. Because so many users fancy themselves experts in user-interface design—or at least qualified to make basic judgments and comparisons—these pressures occur quite frequently. All too often, though, the suggestions conflict, or are simply bad ideas. Under pressure, programmers will sometimes make these changes, only to hear those who exerted the pressure admit, "Well, I guess that wasn't such a good idea after all." When the political pressure is high enough, demonstrating the mistake by implementing it may be necessary, but in general this should be avoided because it is so time-consuming and expensive. (Of course, when good tools for rapid prototyping are available, the trade-off can be very different. If you can build what the users asked for in a very short time, that is often the best way to demonstrate that the users don't really want what they requested.)

Does this mean that a large amount of feedback from the users should really be ignored? Almost. The good designer should keep even the stupidest of suggestions in the back of his mind. If they start to come in frequently, or from more than one source, some serious reflection is called for. First, it is always possible that it is the designer who is being stupid rather than the user; this possibility should never be discounted. More often, the designer may find that even though the suggestions are unreasonable, they point to an underlying problem that was not fully understood.

It may help to think of the user community as being like a preschool full of screaming three-year-olds. One doesn't have to rush to respond every time one of them cries a little bit, as crying is entirely natural for young children. But if some or all of the children begin to wail frequently, something is probably wrong, and an investigation is warranted. If what they're all crying is "I want a cookie,'' that doesn't necessarily mean you should give them all cookies, but you might consider making them a healthy lunch to meet the underlying real need.

Alternately, we can think of users as being like patients, with the interface designers cast in the role of doctors. When someone comes in and says, ''My throat hurts—I think I have throat cancer,'' the doctor need not accept the self-diagnosis, but generally gives credence to the patient's claim to be in pain. Maybe the patient really does have cancer, but more likely it is just a sore throat, easily treated and corrected.

An interesting example of this process can be seen in the history of the Andrew Messages application, a program for reading and sending electronic mail and bulletin board messages. Messages worked on a very large-scale database, with literally thousands of bulletin boards available to the users through a central file system. Performance with such a large distributed database was always a problem, and efforts were continually made to improve the speed with which the program functioned.

One bright idea that was implemented in an attempt to improve performance was *prefetching*. The idea was that while the user was reading a message on one bulletin board, the next message and the next bulletin board would be simultaneously obtained from the central file system. Thus, instead of making the user wait while the data was obtained from the file system, the wait would generally be ''hidden'' during the time the user was actually reading what was on the screen.

That was the theory, at least. In practice, it took a long time to get it right. As it turned out, some of the mechanisms being used to get the data from the file system asynchronously—that is, at the same time as the user did something else—were being unexpectedly serialized at another point in the system. The result was that, instead of making people wait for a file when they actually wanted it, which was the old behavior, the system was frequently making people wait *while the prefetch was going on*!

Users found the process utterly maddening. ''All I want to do is scroll this message forward,'' one said, ''and this stupid program has to go off and fetch some totally unrelated bulletin board before it will let me scroll anything!'' Because of the unexpected wait in the file system code, the actual prefetch system call, which was supposed to take no more than a few milliseconds, was taking as long as five, ten, or occasionally even thirty seconds in practice.

Compounding the problem was the fact that we were slow to understand or correct it. The system performed very differently in different environments, and in this case the problem was far more acute for the users than for the developers. (This is one of the dangers of giving developers particularly powerful machines, although this is often necessary to support development tools such as debuggers.) Once partially diagnosed, the problem proved unexpectedly difficult to fix properly; it was probably half a year after its introduction before prefetching finally started to work the way it was supposed to.

During that time, many of the users became quite certain that we were blithering idiots. The *obvious* solution, to these people, was simply to turn off the prefetching feature altogether, or to allow an option by which individual users could turn it off. Unfortunately, we felt certain that if we did that, we would lose all hope of being able to figure out the problem, because it depended on a complex set of circumstances that we didn't understand and couldn't reproduce. Thus we left prefetching in and studied its behavior until a solution was found. (Unfortunately we didn't study it sufficiently intensively, which is one of the reasons it took so long. This was clearly a strategic error, given the level of dissatisfaction of our users.)

During those months, we were listening carefully to our users' complaints, but were studiously ignoring their suggestions. The users were absolutely right in complaining that there was a problem, but were wrong in thinking that they knew the right solution. There was no convincing them, with arguments, that prefetching was a good thing, because they could see for themselves that, in their daily lives, it was a bad thing. Most were not sufficiently technically minded to follow the arguments at all; they just knew that when Messages began to prefetch something, the whole program froze for a while. Frustrated and furious, they began to talk about their fundamental "right" to turn off prefetching, a concept that made about as much sense as a hypothetical "right" to make us use a particular sorting algorithm instead of a more efficient one.

For the implementors, however, the situation was equally maddening. As computer scientists, we had no doubt that prefetching was a very good idea, and all the screaming was not likely to persuade us that everything we knew was wrong. Nor, for that matter, was it helping us to find the flaw in this particular program.

In the end, when the prefetching behavior was finally corrected, the complaints dried up overnight. While dozens of requests had been coming in weekly to eliminate prefetching or make it optional, not a single such suggestion was ever heard again.[1]   By paying attention to the users' complaints but ignoring their suggested remedies, we were ultimately able to make them far happier, so far as performance was concerned, than they ever could have been had we taken their advice and simply eliminated prefetching altogether.

---

[1]In an odd and ironic postscript to the story, we finally ended up adding a "do not prefetch" option about a year later, when a new version of the file system caused prefetching problems to return for several months. This time, the problem was well understood, but simply couldn't be fixed immediately for obscure technical reasons. In this case, taking pity on our users was harmless, as it would not prevent us from locating and fixing the bug. However, this event had nothing to do with the problems and complaints described in this anecdote, and came considerably later.

```
┌─────────────────────────────────────────────────────────────┐
│ ☐ ≡≡≡≡≡≡≡≡≡≡ Catastrophic Error ≡≡≡≡≡≡≡≡≡≡≡≡≡≡ │
├─────────────────────────────────────────────────────────────┤
│                                                               │
│  An unanticipated error has destroyed all your files.        │
│                                                               │
│       Please click to continue:        [ OK ]                │
│                                                               │
└─────────────────────────────────────────────────────────────┘
```

When something happens about which the users need to be informed, it is generally good practice to ask them to confirm that they have read the message before erasing it from the screen. The Macintosh, along with many other computers, usually asks the user to click on an "OK" button in such instances. While this is not usually a problem, in certain situations it merely adds insult to injury, as in the not-entirely-fanciful example above.

The technique can also lose its value through overuse. Ralph Hill relates the following story: "They [a European computer company] have this wonderful system with a carefully designed user interface that follows all the guidelines. But, no matter how hard they try, they cannot stop the users from executing [some disastrous command] by accident. I ask, 'Do you have to confirm [the disastrous command]?' 'Oh, yes, absolutely. You have to confirm every command.' 'Even simple, safe ones?' 'Of course.' " By requiring the users to confirm every command, they lost the value of asking for confirmation in the dangerous cases; users got in the habit of confirming their actions automatically, and the confirmation lost all its value.

Chapter 14

Lie to Your Managers

When we think we lead we most are led.

—Byron

We have seen that user-interface designers must be extremely independent, in the sense that they must sometimes seem to ignore the expressed wishes and complaints of the very community of users they are supposed to be serving. This independence is presumably justified by the designers' expertise—they should simply know more about user interfaces than the users do, and thus be able to see certain kinds of flaws in their thinking. In this sense the user-interface designer is no different from any other professional except that, because of the nature of the decisions being made, many users consider themselves qualified to advise the designer on his work.

But there is a special class of users who may give the designer advice with an added weight of pressure to *take* the advice. These users are the managers who outrank the designer organizationally *and* actually use his or her software. This is a dangerous combination, for there is no a priori reason to assume that a manager is any better able than anyone else to make decisions regarding user interfaces. Yet, the suggestions of one's manager are difficult to disregard.

Indeed, as with suggestions from any other users, they should be carefully scrutinized for meanings both apparent and hidden. But except for the rare cases where the manager has risen to his or her current position because of (or in spite of) a certain expertise in user-interface matters, there is no reason to assume that the user-interface ideas offered by a manager are any better than those offered by a clerk or secretary. The problem is that when your manager is involved, you have to give some sort of accounting.

Solutions to this dilemma vary as widely as the personalities of individual managers. Some managers recognize the delicacy of their position, and are at great pains to state, on matters of user interface, that "this is only a suggestion." Other managers, however, are used to getting their way by fiat, or through bullying, and must often be resisted more covertly.

The bottom line, however, is that *wrongheaded solutions must be resisted at all costs.* The manager's bad idea is more dangerous than anyone else's, because it is so much more likely to find itself permanently embodied in your code. When the manager has a bad idea, some kind of confrontation is often in the best interest of the project. In the most extreme cases, it may actually be necessary to deceive a manager until the project has evolved to the point where his folly will be obvious to him, but this is clearly a risky career move.

In the Andrew Project, for example, we built a simple user interface to the mail system, called CUI. CUI was never intended to have a particularly good user interface. The acronym stood for "Common User Interface," and CUI was intended to be precisely that—the simplest possible interface to the mail system, usable even from a teletype, and available on a wide range of systems. Users desiring a "friendlier" interface were expected to use a different one. CUI was the "last resort" interface, for use on low-end terminals and on platforms to which the better interfaces had not yet been ported.

The problem was that in the earliest releases of the mail system, only two interfaces were available—the primitive CUI and the flagship Messages program, which only ran on high-function bitmap-display workstations. Anyone working, for example, via a dialup line from home, or from an IBM PC, had to use CUI.

Nobody particularly liked this situation, and one manager (fortunately *not* the manager of the mail system group) got it into his head that the problem was that the designers of CUI didn't know how to build a good user interface. He started up a task force on "low-end mail-reading interfaces" and blustered his way through a good many meetings on the topic. Eventually, he drafted a proposal specifying in excruciating detail what a "good" user interface to replace CUI would look like. All in all, he made life very unpleasant for the mail system group, who had never intended CUI for widespread use.

The end of his crusade came rapidly, from two directions at

once. First of all, his draft proposal was widely decried as being "even worse than CUI." This was not surprising, given that the manager had no previous experience with user-interface design. More important, the mail system developers, continuing with the original plan, released a new intermediate interface, much nicer than CUI but requiring at least an intelligent video display terminal. The dissatisfied CUI users now had the long-awaited better alternative, and very little was ever said again about the "need" to improve CUI, which remained the ugly lowest-level interface it had always been intended to be.

The key point is that, like any other technically uninformed user of the software, the manager was able to perceive correctly that a problem existed, but was unable to understand the solution envisioned by the designers. His position as a manager, however, enabled him to make life less pleasant for the designers over a period of time. Indeed, we got through those several weeks largely by dissembling, by pretending to be paying attention to his plans while we proceeded to implement our own rather different ideas instead.[1]

Confrontation with managers can be necessary at any stage in a project. In the early days of the Andrew Project, the original designers quickly concluded that the software should be built for engineering workstations under the UNIX operating system. This came as a surprise to IBM, the funder of the project, which had no such machines at the time and was expecting to make a generous donation of IBM PCs. It was, to say the least, unnerving for them to discover that over a million dollars of their money was to be spent on the purchase of workstations from a competitor, Sun Microsystems. By standing up to IBM on an issue of clear technical importance the Andrew developers not only helped ensure the success of Andrew, they also made IBM more aware of a serious deficiency in its product line, and thus helped motivate the later development of IBM's UNIX workstations. It is frequently good even for the *managers* when you diplomatically confront them about their mistakes.

---

[1]The reader should not feel sorry for the poor humiliated manager. He landed catlike on his feet, claiming publicly that the only reason the new interface existed was the pressure he had exerted on the developers. Getting the last word is generally a manager's prerogative, and, at any rate, it is often important to leave some way for them to save face and take credit for the final product.

```
$ is
Welcome to the [xxx] system, version 4.32.a.
This program was written by [xxx]
in the Schools of Engineering at [xxx] University.
$ Would you like to submit a query? y
Valid answers are yes and no.

$ is
Welcome to the [xxx] system, version 4.32.a.
This program was written by [xxx]
in the Schools of Engineering at [xxx] University.
$ Would you like to submit a query? yes
What language would you like to type your query in? french
Sorry, only english is currently implemented.

$ is
Welcome to the [xxx] system, version 4.32.a.
This program was written by [xxx]
in the Schools of Engineering at [xxx] University.
$ Would you like to submit a query? yes
What language would you like to type your query in? english
Type in your first query word: computer
Would you like to type another query word? n
Valid answers are yes and no.

$ is
Welcome to the [xxx] system, version 4.32.a.
This program was written by [xxx]
in the Schools of Engineering at [xxx] University.
$ Would you like to submit a query? yes
What language would you like to type your query in? english
Type in your first query word: computer
Would you like to type another query word? no
Would you like to evaluate the query? yes
Query evaluated.  Would you like hard copy? yes
Hard copy not available.
$
```

Some programmers seem to equate "friendliness" with verbosity. This is a mistake, as the above dialog demonstrates. (based on an actual database system, reported by Scott Deerwester)

Chapter 15

Cut Corners Proudly

He is not only idle who does nothing, but he is idle who
might be better employed.

—Socrates

Tests were conducted through the evening to see whether
the glue would be dry in time for liftoff.
    —from a news report on the successful launch of
the Space Shuttle Columbia on June 5, 1991

A good professional takes pride in his work. A master carpenter tries
to make every wall meet the floor, ceiling, and other walls at perfect
right angles, even where nobody is ever likely to notice or care. To
an electrician, a neatly wired circuit can be a thing of beauty, far
beyond the requirements of safety.

Programmers, too, have a sense of beauty in their work, and
strive for cleanliness, elegance, and simplicity. To a professional
programmer, such things are points of honor. A program that works,
but is ugly and confusing in its implementation, is derided by any of
several terms from the programmer's argot: *crock*, *kludge*, *hack*, or
several less printable phrases. A professional programmer, it is well
known, should expend the necessary time and effort to produce code
that he can be proud of, code that other programmers can easily read
and understand.

Historically, the development of a professional programming
culture that promotes high-quality code was a great thing. In the early
days of programming, any program that worked was a good program.

The result, people quickly discovered, was a proliferation of "unmaintainable" programs that worked for a while but ultimately could not be fixed or upgraded when necessary, and eventually had to be scrapped. "Quick-and-dirty" programs began to be seen as unprofessional, *primarily because they were not cost-effective*. Short "throwaway" programs that were intended to be run only once or a few times, perhaps to answer some simple question, were still somewhat acceptable to write in the old quick and careless style because they would not need to be maintained. But even there, extreme sloppiness was discouraged because of the surprising frequency with which "throwaway" programs ended up living for years.

In general, however, most programs were not written to be thrown away—which was good, because most programs were *not* thrown away. Most programs were written after careful specification and design, and were intended to last for many years of service (often many thousands of machine-years of service, given the number of computers the software would ultimately run on). The programmer's notion of professionalism is explicitly geared toward the notion of programming "for the ages."

As with so much of professional software engineering, however, this "professional pride" is often misplaced and wasted when applied to developing user-interface technology. This is because one of the basic assumptions of that pride—namely, the idea that the code is to last a long time—is so often violated. Certainly user interfaces are often used for many years, but the way they develop is different.

User interfaces are fundamentally *evolutionary* artifacts, even more than other computer programs. No matter how carefully designed they are, they typically change frequently, often radically, in the early part of their lifetimes. Thus the effort to make the code "perfect" in the early stages of user-interface development can be essentially wasted, as the purity of the code is quickly sullied by the attempt to make the interface more pleasant. Moreover, prototyping sometimes even reveals that whole features originally planned for the interface are unnecessary, and can be omitted from the final product. It makes little sense, if this is going to happen, to put any more effort into the prototype than is absolutely necessary.

Ultimately, the problem is that user interfaces are much harder to design in advance than other programs, and so it makes less sense to approach them "professionally" and write them to last for decades.

A better approach, in many cases, is to make the initial user interface a quick-and-dirty component of the larger system, the rest of which is built as a more stable and lasting edifice. A clean separation between the interface and the rest of the system will aid in this process enormously, as the other parts more typically *are* understood well enough to be worth writing "properly." Then, once the interface has stabilized and is generally well-enough liked, it can and should (if it seems appropriate and cost-effective) be rewritten in the traditional style of professional programming.

Professional Prototyping Practices

The essential notion here is *prototyping*. Every new user interface is, of necessity, a prototype. Almost inevitably, a good user interface is the product of the evolution of many different prototypes, or at least many different versions of the prototype. The ideal process of creating a good user-interface product, then, is one of producing repeated prototypes until the result is deemed good enough, and then reimplementing that final prototype as a lasting product. This process is actually fairly well recognized and understood among experienced developers of user interfaces.

However, its implications are less well understood, particularly by software engineers and managers. There are enormous differences between prototype software and production-quality software, and the difference is not merely in how unreliable or "flakey" prototypes are. They also reflect different goals during the process of producing the software.

In prototyping, the goal is simply to get something running. It doesn't matter if the program happens to die every time you hit the F1 key, or if an obscure sequence of commands puts it in an infinite loop. Those are trivialities compared to the larger goal of getting the system running in a form that can be tested on the users. For a production system, on the other hand, those kinds of problems are usually utterly unacceptable.

If an enlightened approach to user-interface development is to develop repeated prototypes until one is satisfactory, and then to reimplement it rigorously, this means that most of the user-interface development should take place as a prototyping activity. This implies, practically speaking, that programmers must learn to disregard their notions of "professional pride" and produce what they have typically thought of as amateurish code, quick-and-dirty hacks and kludges that absolutely will not stand the test of time. In developing these prototypes, programmers should cut corners whenever possible, skimping on reliability, robustness, and efficiency in order to get their prototypes up to a level where they can be tested on real users. (Of course, a certain minimum level of reliability and efficiency is necessary or the users will notice, and be unable to cope, even in the limited situation of testing and evaluating a user interface.)

In general, the corners one can cut involve robustness and program behavior in rare circumstances. It is generally acceptable for prototypes simply to die when they get highly unusual input, when they run out of memory, or when an underlying subsystem fails. It is probably acceptable if a prototype depends on system features that are being phased out in future releases, or if a prototype incorporates algorithms that are known to be inefficient in some circumstances. It is generally vital, however, that such shortcuts be avoided or rectified in the final product. Thus the real problem, for developers, is how to make sure that the shortcuts in their prototypes do not persist in the final product.

Discarding the traditional standards of software engineering professionalism does not mean that "anything goes" for the prototype builder. What it means, in practice, is that a different set of standards apply. In particular, it is important to remember, as you fearlessly cut corners, that if the final interface ends up looking anything like the current prototype, someone will probably have to come back and "uncut" all of these corners. (That "somebody" might even be the person who built the prototype, but more likely it will be someone else, since individual temperament seems to make some programmers better at prototyping and others better at building "permanent" software.) This means that it is the professional responsibility of the prototype builder to *document* all the corners he cuts.

This, for traditional programmers, is a horrifying prospect. All programmers cut corners sometimes, but this is like saying that all men are sinners—it may be true, but it doesn't imply that all men keep a careful list of their transgressions to show to anyone who asks about them. Programmers are used to feeling guilty about cutting corners, and thus tend to bury these shortcuts where others are unlikely to find out about them. For the programmer building a prototype user interface, however, it isn't shortcuts that are unprofessional, it is hiding them, or indeed failing to make them obvious. A "good" prototype will be full of shortcuts, but each will be briefly documented on a list of things that need to be cleaned up, rewritten, or redesigned for the final version of the product.

The existence of such a list—an extreme rarity—can redefine the relationship between the prototype builder and the rest of the product developers. Instead of being the idiots who cut all those corners, the prototype builders can demonstrate their basic professional competence and understanding of software engineering by pointing out all the corner cutting themselves. The people responsible for the final product will also be able to build it more quickly if they are armed with a fairly complete and detailed list of the problems of the prototype. In particular, such a list will greatly simplify the choice between a complete rewrite and a thorough cleanup of the prototype.

The user-interface designer, then, should cut corners fearlessly and proudly, but should keep careful note of everything he or she does wrong, so that if the interface happens to turn out to be worth using, it can be rewritten properly.

Occasional Corners Can Stay Cut

Very rarely, this approach can pay off in unexpected ways, by revealing situations in which "professionalism" may be misguided and unnecessary even in a final product. One unusual example of this is offered by the Andrew Console program, first discussed in chapter 10.

The Console program was customizable through user programming in a programming language designed expressly for the purpose. This language, known as LACC (the Language for Andrew Console Construction) had an extremely simple syntax. When Console was first built, it was unclear how important LACC would be, how often it would be used, or indeed even if Console itself would be a success.

It was, therefore, not appealing to apply traditional techniques to the implementation of LACC. Building a full-fledged interpreter or compiler takes a fair amount of effort, and is hard to justify for an enterprise as unproven as Console. (Indeed, the entire enterprise of building Console was barely tolerated by much of the senior management of the Andrew Project. They saw it as largely a waste of time, so it was essential to get it built quickly, before they ran out of patience with it. Later, of course, with perfect hindsight, the same managers described Console as one of the most vital Andrew applications.)

Instead of building a "proper" language interpreter, a "quick-and-dirty hack" was used for the first version of Console. Instead of implementing the LACC interpreter properly, LACC was designed to be interpreted by a "finite state machine"—a particularly simple program to build—and to *ignore* all of the punctuation and other syntax that were defined as part of the language. What this meant, in practical terms, was that where the "correct" LACC syntax might be something like

```
@instrument(xmin 20, xmax 25,
        ymin 30, ymax 50)
```

one could instead simply write

```
instrument xmin 20 xmax 25
ymin 30 ymax 50
```

or

```
$instrument {
    xmin := 20;
    xmax := 25;
    ymin := 30;
    ymax := 50;
}
```

or even

```
(instrument}>
xmin != 20 +
xmax += 25 !!!
ymin < 30 ~)
ymax >>= 50 {
```

In other words, the interpreter was, by the normal standards of professional programmers, a joke. It didn't recognize the language as specified, but recognized and interpreted a vast superset of the correct language syntax. However, it was extremely easy to implement (it took just an hour or two to write and debug the parser), and it was implemented with the intent to rewrite it "properly" eventually if it proved sufficiently useful—that is, with the intent to come back later and "uncut" the corners.

Gradually, a surprising thing became apparent: there was no reason to rewrite it. LACC was surprisingly popular, in part because it was so easy to write programs in. "Amazingly, my LACC programs often seem to work on the first try!" one user exulted on an electronic bulletin board. The fact that the interpreter recognized (or rather, *ignored*) such a wide range of syntax meant that users could accidentally leave out various forms of punctuation and still have the right thing happen. What was designed as a temporary hack soon came to be perceived as one of the great virtues of the system, and the idea of writing a pickier "proper" LACC parser was quietly abandoned. The language interpreter continued to be offensive to anyone who knew how to write a proper parser, but the advantage of the more liberal interpreter for casual users was undeniable.

Of course, most languages aren't simple enough (or harmless enough in the possible consequences of bugs) to be implemented this way. This story isn't really about a new approach to implementing programming languages. The real point is that sometimes cutting corners can yield unexpected benefits. By keeping the prototype to a bare minimum, one occasionally finds that the bare minimum is closer to being sufficient for the final product than originally expected.

It was my first week on a new systems engineering job and I had to have a revision of a several-hundred-page specification document done before an 8:30 A.M. meeting. I was working late into the night, and at about 2 A.M. I finished the content of the document.

I was using nroff as my word-processing system, with ed as my text editor. I ran the file through nroff and then ran a program called diffmark, which puts marks in the margins to indicate where changes had been made, so that small changes don't get lost. I ran diffmark and then inspected the file before I sent it off to be printed.

Entering ed, I typed "1p" to get ed to print the first line of the document. Ed replied by simply printing "?" alone on a line—ed's standard answer when it doesn't know or can't do what you want. This was scary, because the only reason I could think of for ed to be printing a question mark was that the file was completely empty!

Further exploration seemed to show that ed was extremely broken. The file was very long, and ed would show me portions of the end of it, but when I tried to see anything near the beginning of it—say, the first, tenth, or hundredth line—ed would only print out its single question mark.

$ ed output-file
182
1p
?
10p
?
100p
?

After an hour of increasing panic, I figured out that diffmark had placed 134 lines containing nothing but a single question mark in the front of my file! There was nothing wrong with ed, but there was no way to tell the difference between ed's response when it couldn't do what you wanted and its response when asked to print out a line containing a question mark. (submitted by Bonnie John)

Chapter 16

Remember Your Ignorance

Looking upon myself from the perspective of society, I am an average person. Facing myself intimately, immediately, I regard myself as unique, as exceedingly precious, not to be exchanged for anything else.

No one will live my life for me, no one will think my thoughts for me or dream my dreams.

In the eyes of the world, I am an average man. But to my heart I am not an average man. To my heart I am of great moment. The challenge I face is how to actualize the quiet eminence of my being.

—Abraham Joshua Heschel

It is never too late to give up our prejudices. No way of thinking or doing, however ancient, can be trusted without proof. What every body echoes or in silence passes by as true today may turn out to be falsehood tomorrow, mere smoke of opinion, which some had trusted for a cloud that would sprinkle fertilizing rain on their fields. What old people say you cannot do you try and find that you can. Old deeds for old people, and new deeds for new.

—Henry David Thoreau

Designing and building a large program is a complicated business. In a sense, nobody can do it *right*. As with any other human artifact, a computer program is inevitably filled with small flaws that reflect the inadequacies and failings of those who built it. The goal, of course, is to minimize and hide these flaws to the greatest extent possible.

Vital in doing that is the recognition of one's own limitations. Programming can be an exciting, intoxicating activity. Non-programmers rarely understand that the underlying cause of the obsessive programmer's fixation on his or her work is, in essence, megalomania. In programming, one creates a world of one's own, and is limited within that world only by one's own vision and talent. Success in that world feeds programmers' sense of their own omnipotence—particularly heady stuff for the adolescent mind. Moreover, computers are seductive. They are nearly perfect servants, undemanding when ignored, attentive to a fault when we focus on them, and nearly always at our beck and call.

Ultimately, of course, we are *not* omnipotent, and it is against the unforgiving rocks of this truth that many promising software expeditions have been shipwrecked. Few things are harder than to persuade an excited young programmer to ask someone else for help. Yet such help is often as easily available as it is sorely needed. The problem of finding help is almost insignificant in comparison with the problem of realizing when it is needed.

For the user-interface programmer, many kinds of specialized experts are readily available to offer help. Wherever words are involved, for example, writers may be useful. The casual message to the user, which the programmer dashes off in order to be able to test the latest behavior of the program ("illegal command—aborted"), can be clarified and amplified by a writer who understands what is going on ("The command 'foobar' is not meaningful and will be ignored").

Similarly, the visual display presented by a program can nearly always be improved by an artist, perhaps a graphic designer, whose expertise and training lie precisely in the area of making things pleasing and comprehensible to the human eye. James Gosling, the designer of the original window management system used in the Andrew Project, described his own first experience with the graphic designer who served as a consultant in that project:

> He spent about a week playing with the program, testing things, staring at it, and so on. I'd walk by his office, and he'd be staring intently at the screen, not moving except occasionally to write something down on a piece of paper. I was afraid that, in the end, he'd come to me with some wild, outrageous set of major changes to make. When he came

back to me, though, what he finally said was that the gray area that separates the windows was too small, and that I should enlarge it by a few more pixels. It seemed like an incredibly trivial result after all that work, but I tried it, and it actually did look a lot nicer that way.

In this case, the programmer found the artist's methods completely inscrutable, but was sufficiently open-minded to recognize the value of the results, and to make appropriate use of them.

Once the right mind-set is obtained, the programmer can find useful advice in a wide variety of places. By recognizing one's own limitations and broad state of ignorance, one can even find times when the opinions of other programmers can be useful. Often another programmer can offer specialized advice on the selection of algorithms or the design of subcomponents (e.g., small languages and databases) that will lead to a more efficient or more flexible system.

Writers are useful, artists are useful, and even other programmers can be useful to the interface designer. Psychologists or educators can be useful in figuring out how to present particularly complicated material to people. As the multimedia capabilities of computers increase, a variety of new experts, such as musicians and cinematographers, may likewise come to be useful. Software is such flexible stuff that nearly any field might be expected to provide useful advice some day. One never really knows in advance which kind of advice is going to prove to be the most important. Keeping an open mind, therefore, can lead to the most wonderful surprises.

For example, many people have, over the years, built programs known as *online help systems*. These are simply programs that answer a user's request for help by providing him with information on how to use the larger software and hardware system of which they are a part. For my doctoral research (Borenstein 1985a), I decided to conduct some controlled experiments on such systems, in an attempt to get some actual data that would help determine the "right" way to provide online help. I videotaped dozens of subjects as they tried to perform a complex set of tasks, the only variables in the experiments being which help system they were given to use and which set of help texts the system would show them.

Much to my surprise, the experimental results showed that the user interfaces by which help was obtained, though they varied greatly in quality, were almost irrelevant in determining how fast the users figured out what they wanted to know. Far more important, it turned out, was the quality of the help texts actually presented to the users. That is, having a good writer producing well-written help *texts* was actually far more important than having a good programmer producing a snazzy help *system*.

The real moral of the story, however, was not in the experimental results themselves, but in the way they were received. Naturally, the results were welcomed by technical writers, who have used them to help sell their services to software and hardware vendors. But although they welcomed my results, few writers were surprised by them. "Of course the text matters more," one said. "It's great to see it verified by experimentation, but it's hardly surprising." In the world of computer science, however, the reaction was rather different. For the most part, the result *was* surprising to programmers. It was, for some, a rare inkling that the programming might not be the most important part of a software system.

It shouldn't have surprised anyone, but it did. It certainly surprised me. If I'd been smarter, though, I would have consulted some technical writers at the earliest possible stages of my study of online help systems. They probably would have told me then, before I did any experimentation, that the quality of the texts was the single most important factor. Unbeknownst to either me or them, they were already, by virtue of their specialized training, more expert in some aspects of the area I was studying than *I* was. Such ignorance is very common among programmers, and is a great handicap to the production of high-quality interfaces and applications.

In general, programmers are highly specialized experts. When they build user interfaces, they nearly always sail into uncharted regions of which they are deeply ignorant. The best thing they can do, in such circumstances, is to recognize their ignorance as fully as possible, and to consult the people who know more than they do about the matters at hand.

It should be noted, however, that the experts cannot be trusted blindly. Experts are wrong sometimes too, and their advice must be carefully weighed and evaluated, both before and after one takes it. I have on a few occasions taken the advice of graphic designers regarding improvements to the visual display of a program, only to have to withdraw the "improvements" after the users howled in outrage.   In one memorable case, the graphic designers suggested making a screen display "easier on the eyes" by changing the most important text region in the Andrew mail-reading program from single to double spacing.   The users quickly discovered that the drawback of having only half as much information in view was much more significant than any gain in aesthetic appeal, and they told us so in no uncertain terms.   In this case, the graphic designers were probably right about what was prettiest, but "pretty" was not what the users cared about most in this case.   The lesson here is that, even as you remember your own ignorance and fallibility, you must remember that the experts who advise you are comparably, but not identically, limited in their vision.

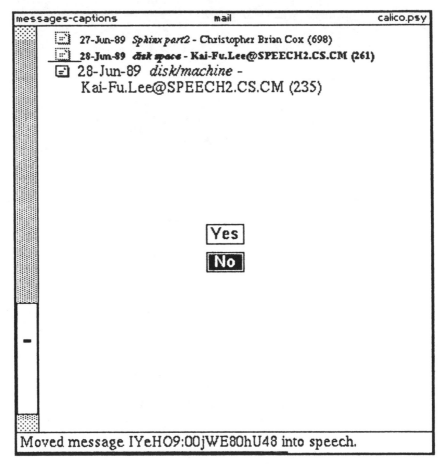

27-Jun-89 *Sphinx part2* - Christopher Brian Cox (698)
28-Jun-89 *disk space* - Kai-Fu.Lee@SPEECH2.CS.CM (261)
28-Jun-89 *disk/machine* -
Kai-Fu.Lee@SPEECH2.CS.CM (235)

Yes
No

Moved message IYeHO9:00jWE80hU48 into speech.

A dialog box is a very common mechanism with which to ask users questions or inform them of important information. Bugs in the actual dialog box code are, therefore, particularly nasty and troubling to users. In this case, both the question and the box itself have mysteriously failed to appear, leaving the user with a choice that resembles nothing so much as a Zen koan.

Chapter 17

Dabble in Mysticism

A good traveler has no fixed plans
and is not intent upon arriving.
A good artist lets his intuition
lead him wherever it wants.
A good scientist has freed himself of concepts
and keeps his mind open to what is.
Thus the Master is available to all people
and doesn't reject anyone.
He is ready to use all situations
and doesn't waste anything.
This is called embodying the light.
What is a good man but a bad man's teacher?
What is a bad man but a good man's job?
If you don't understand this, you will get lost,
however intelligent you are.
It is the great secret.

—Lao Tzu, *Tao Te Ching*

Good advice about user interfaces can come from almost anywhere. Professional arrogance is a major enemy of user-interface programmers, because it can blind them to good ideas from "unprofessional" sources. A better attitude might be to imagine that good ideas are constantly surrounding us all the time, in a sort of ether-of-ideas. Through unimaginable mystical processes, these ideas may make themselves known to us through the comments of a fool, or even through a Ouija board. A constantly open mind is the designer's most important tool.

In these early years of the Computer Age, programmers are naturally filled with a sense of self-importance. We are actively defining a new world of human activity—at least if one believes, as many do, that programming is unlike anything that has come before. It is easy to think of programmers as a special breed, without need of the assistance or advice of the primitive humans who have no idea of the joy and challenges of computer programming.

This self-importance, though overstated, is not always misplaced. In general, programmers work on projects in which their expertise is paramount.   They are the true experts on software efficiency and reliability, for example.   In certain applications, consultation is required—astronomers can be useful "consultants" if you're writing programs to control rocket ships—but the need for such specialized experts does little to disabuse the programmer of the notion that he is largely independent of outside help.

In user interfaces, however, this notion is just plain wrong. Traditional programming concerns like speed and robustness are still issues, of course, and the programmer is the right person to solve them. But a host of other concerns are also relevant to user-interface design, most of them outside the programmer's expertise and training.

Some of these concerns are narrow and almost technical in nature. As mentioned before, the visual appearance of a program on the screen is likely to benefit by the consultation of a graphic designer. Similarly, the wording of explanatory texts and dialogs can be greatly improved by the efforts of a good technical writer. Although many programmers never think of getting graphic designers or writers to consult with them, most will, upon consideration, readily accept this idea as being similar in spirit to having astronomers consult on spaceship control software.

Thus, like almost any other kind of software, user-interface software can benefit substantially from the consultation of relevant experts. However, there is a unique kind of expertise that may be beneficially applied to user-interface software, but that is harder for most programmers to accept or understand.   Often the best suggestions will come from the *real* experts—the users themselves.

This is less distressing in principle than it is in actual practice. In principle, giving thoughtful consideration to advice from the real users of a system is a wonderful idea, hardly subject to debate. In practice, however, it is often a nightmare, largely because there are usually so many users and their needs, and their perceptions of their needs, and their ability to articulate their needs, vary so greatly. In order to take the advice of users without being overwhelmed or intimidated by it, the programmer must become something like a physician in combat, performing triage to separate the frivolous and the hopeless from the urgent, and performing instant analysis to figure out who is truly sick and who is merely malingering, or what the users *really* want when what they *say* they want is patently absurd.

After a while, with a stable user community, the job can become easier as the programmer learns which users can be counted on for sane, useful suggestions, and which can be expected to submit an endless series of impractical ideas or time-wasting minor suggestions. It helps to know that some people fit each of these categories, but every user comment still deserves consideration.

In my own experience, the best user suggestions seem to come from sophisticated nonprogrammers. Within a development staff, for example, one can often find the most critically adept users in the documentation staff, the testing staff, or the support staff (secretaries, for example). Just as the ability to write a good book and the ability to be a good book reviewer seem to be largely independent, so, too, the ability to make useful suggestions about a user interface seems to be independent of the ability to build one.

Feedback from users is one of the most critical requirements of successful user-interface development. The prototypes must be given to the users at the earliest possible moment, and simple channels for feedback from the users must be created. Electronic mail and bulletin boards seem tailor-made for this purpose, but where they are not feasible, other alternatives must be found, such as regularly distributing evaluation forms to the users.

It is sometimes hard for programmers to accept the notion that it is worth listening to the advice of people with absolutely no relevant technical expertise. Taking the advice of specialists is not as hard because programmers understand well the notion of specialization. Taking the advice of "random idiots off the street" is a little harder to swallow, but is ultimately essential if you want these same "random idiots" to be able to use your program when it is done. A slightly mystical attitude regarding the nature and sources of good ideas may not be essential, but it does seem to help. Good ideas can appear from anywhere, at any time, and when inspiration materializes out of the apparently empty head of a particularly annoying user, it may be hard *not* to view it as a mystical event.

Bravo was a text editor that ran on the Xerox Alto computer, a computer that is widely recognized as the ancestor of the Macintosh and other modern "friendly" user interfaces. The Alto pioneered the use of high-resolution displays and pointing devices, among other things. Bravo was, in many senses, the first text editor with a "modern" user interface.

Bravo had, however, a rather amazing flaw in its user interface. Like many older editors, the interface functioned in two modes—a "command" mode, in which all user input was interpreted as commands, and an "input" mode, in which the user typed text to be inserted into a document. The problem occurred if a user happened to type the word "edit" while in command mode. Bravo interpreted the "e" to mean "select everything." The "d" meant "delete the currently selected item." The next letter, "i," put you back into input mode. The final "t," therefore, was the text to be inserted. Thus typing "edit" was interpreted to mean "delete my whole document and replace it with a single letter 't.' " While Bravo did have an "undo" facility, it was only able to undo the most recent operation. Thus the final "t" ensured that the text that had been deleted could never be retrieved, because its deletion was no longer the most recent action.

Chapter 18

Break All the Rules

Even if we achieve gigantic successes in our work, there is no reason whatsoever to feel conceited and arrogant. Modesty helps one to go forward, whereas conceit makes one lag behind. This is a truth we must always bear in mind.

—Mao Tse-tung

The only reasonable eleventh commandment is that you shouldn't take the other ten too seriously. Although they're generally sensible and worth paying attention to, and although one can come up with many good examples of the relevance of each, none of these commandments were, in fact, revealed to mankind by a deity who demands literal observance. Each may be considered provisionally true, but counterexamples can always be found. In the final analysis, it is the good judgment of the individual designer that makes all the difference. Rules such as those given here can help inform that judgment, and can perhaps help designers to develop their own ability to make such judgments, but they can never replace the need for the designer to think and decide.

At some level, the user-interface designer must, inevitably, be an artist. This is not, by any means, a stretching of the term ''artist'' to suit the modern sensibility. The *Oxford English Dictionary* provides four major definitions of an ''artist'':

1. *One skilled in the ''liberal'' or learned arts.* We have seen that a good user-interface designer uses ideas,

skills, and advice from a wide variety of specialized fields, from the aesthetic advice of a graphic designer to the technical advice of the electrical engineer. Such interdisciplinary activity is the hallmark of the liberal arts.

2. *One skilled in the useful arts.* A user-interface designer is, primarily, a programmer—whose craft is at least as useful as those of the tailor or potter, traditional examples of the "useful" arts.

3. *One who pursues an art which has as its aim to please.* Although this is not a primary aim of most programming, it is in fact crucial to user-interface programming. A good interface gives at least a subtle, quiet pleasure to its users, by making minimal unnecessary demands on their skill and memory. Additionally, visually attractive or elegantly structured programs can certainly please their users aesthetically.

4. *One who practices artifice.* "Artifice" is another term that is defined at length, but a crucial element in several of the definitions is cunning or strategy. The artist *fools* the public with his art. A good user interface will often create metaphors, such as the "desktop" metaphor on the Macintosh and other computers, which have little or nothing to do with the underlying reality of the computer. Such metaphors fool the user into thinking with an incorrect but easily understood and consistent model of how things work. Good interfaces also hide the vast complexity of the underlying systems, another example of software artifice.

Moreover, user-interface design is not merely an art; at its best it is a *performing* art. In an ideal environment, in which designers get frequent and rapid feedback from their users, the designers are always on stage. Every time users sit down to use a program, it is as if the designers themselves are standing in front of them, juggling flaming swords. Every time the designers sit down to improve their programs, they should be able to feel a crowd of users watching over their shoulders, waiting to see what they create.

In fact, the more designers can persuade their users to think of them as artists and performers, the better off they'll be. People have, for some reason, a tremendous tolerance for artists. They offer them a great deal of freedom for both professional and personal idiosyncrasies. The general respect that people hold for artists—which, of course, is only rarely translated into financial terms—means that they presume less to tell them what to do. While advice and feedback from users is essential, orders and requirements from managers or important clients is, as discussed in chapter 14, often destructive. If the managers and clients think of the designer as an artist, they are less likely to turn *suggestions* into *requirements*.

The user-interface designer *as artist* will not, in most cases, be judged by his adherence to formal requirements or design methodologies. Indeed, he will not even be judged by his adherence to a particular philosophy of user-interface design, such as that found in this book. He will, in the end, be judged only by the quality of the interface he has managed to produce, its suitability for use by the widest possible range of users, and the amount of pleasure that people will take from it when they use it.

Nothing else matters. If the interface is good enough, almost anything else will be forgiven. Certainly badly engineered, unmaintainable software is forgiven as a matter of routine. Even a screen display that is "ugly" by most people's standards will be forgiven if it fits the product in such a way that it doesn't give offense to the people who finally use it. Every rule can be broken for the sake of the final product, though the more rules you break, the worse you'll look if the final product is *not* successful.

In the end, if you keep the users happy, you can get away with murder. They may not even complain if you quietly slip in something like an eleventh commandment in a set of ten.

Part Four

The Golden Path:
The Road to Human-Oriented
Software Engineering

Just before Ninakawa passed away the Zen master Ikkyu visited him. "Shall I lead you on?" Ikkyu asked.

Ninakawa replied: "I came here alone and I go alone. What help could you be to me?"

Ikkyu answered: "If you think you really come and go, that is your delusion. Let me show you the path on which there is no coming and no going."

With his words, Ikkyu had revealed the path so clearly that Ninakawa smiled and passed away.

—*Zen Flesh, Zen Bones*

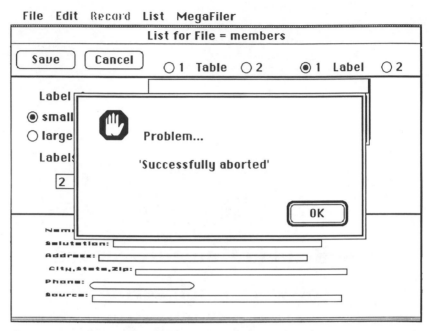

Sometimes the most obvious situations and solutions are simply overlooked. Here, the designers of a user interface failed to anticipate the very likely situation of a user cancelling an ongoing print job, with the amusing result shown above.

In another example of overlooking the obvious, the computer facility for a state university had a time-sharing system, with several interactive editors. One of the editors did not force the user to write the file before exiting. A high official solved the problem by decreeing that all programmers had to follow a formal checklist to save files and exit the editor. This reduced the errors but did not eliminate them.

Unnoticed during all this decree making was the fact that another locally available editor had a similar syntax to the favorite editor, was more powerful, ran faster, and forced users to save their files before exiting (or use a special command to exit without saving their files). People were too busy trying to "fix" the old one even to notice that a better alternative was available. (suggested by Ralph Hill)

Chapter 19

The Tools of the Trade

Learning is a movement not anchored in knowledge. If it is
anchored it is not a movement. The machine, the computer,
is anchored. That is the basic difference between man and
the machine. Learning is watching, seeing. If you see from
accumulated knowledge then the seeing is limited and there
is no new thing in the seeing.

—Krishnamurti

In most of this book, I have argued that user-interface design is so
idiosyncratic, so unpredictable, and so fundamentally *not* an
engineering discipline that I may have given the impression that
software engineering has nothing to offer user-interface designers.
This is an exaggerated perspective, offered in frustration by one who
has been trained in "traditional" software engineering and found it
wanting for his purposes. This does not imply, however, that it is
utterly valueless—merely that it is only one domain among many that
must be mastered by the user-interface designer.

In fact, designers are not completely helpless and unarmed in
their crusade to build good user-interface software. Over the years, a
sizable arsenal has been built up for the software engineering "arms
race," and user-interface designers should take full advantage of it.
Much of this arsenal has been assimilated as "common sense" by
good programmers everywhere, and would hardly deserve mention
were it not for the fact that so many user-interface designers come
from backgrounds other than those of the traditional programmer.

In order to be a *good* user-interface designer, one must

understand not only all of the unique truths of user-interface programming, but also the more standard truths of general software engineering. One must also avoid being tricked into unwarranted reliance on some of the unproven techniques for which grandiose claims are often made. In this chapter, we will peek into the software engineer's bag of tricks and try to see what of value is hiding inside it.

Standards

Businesspeople, as mentioned in chapter 6, love standards. Without standards, product developers take much bigger gambles with every innovation. Had Sony and its competitors agreed on a single format for videotapes before the great VHS-Beta war began, the other companies would probably have made no less profit, but Sony's position would have been much better.

The discussion of standards in chapter 6 was essentially negative. Standards are, from the user-interface designer's perspective, inherently constraining. Nonetheless, standards are necessary for other reasons, and once they exist they can be very useful to the designer. Principally, the designer can use them as a guide to determine what decisions require thought and which are automatic. Designers don't have to like a company's standards for software "look and feel" to realize that there are some aspects of the design on which they needn't waste their time.

Moreover, standards relating to graphical interfaces and data interchange are a clear boon to the interface designer, because they make software more portable and more compatible with other programs. The alert designer will always be sensitive to emerging standards, and will try to avoid getting caught with incompatible software. Never assume that newly defined standards are irrelevant to your own work—the consequences are too enormous when the assumption is erroneous.

In almost any part of the computer industry, one can find both good and bad standards. The really bad standards are usually replaced eventually, but the process can be slow and very painful for all concerned. (Consider, for example, the slow progress of high-definition television pictures in the United States, which is essentially the process of devising a replacement for the old NTSC video standard, widely viewed as a particularly bad standard.) Technical considerations *can* derail a de facto standard, however. Gradually, systems with enormous market weight like CP/M and MS-DOS appear to be giving way to more technically reasonable standards like UNIX. Would-be standards with serious problems can fail even with enormous market resources supporting them. The Ada programming language (which has been backed to the hilt by the Department of Defense, but has yet to win over many converts among those who aren't *required* to use it) may yet prove to be an example of such an unsuccessful standard.

Methodologies

Much of standard software engineering research is devoted to the topic of "methodologies." Methodologies are simply organized techniques for moving methodically toward a goal. In the case of software engineering, methodologies are primarily organizational techniques by which the various steps of the software development process are planned to result in the best possible final product.

Unfortunately, most of the standard methodologies were designed for software other than user interfaces, and are strikingly inappropriate for the intensive prototype-test-revise cycle that is characteristic of such software. Nonetheless, these methodologies may still be useful at the end of that cycle, if the basic interface has been stabilized and a complete reimplementation is desired. Software engineering has traditionally assumed that there would be a clear specification in advance of any code writing, so the established methodologies are well worth considering whenever that situation applies.

Beyond the grand methodologies, however, are the practical ones. There are a number of simple disciplines practiced by good programmers around the world, which user-interface developers should all know about and practice as a matter of course, though they rarely do:

> *Version Control*. Especially in the rapid prototyping phase of user-interface development, good version control is essential if the developers are to have any hope of tracking and fixing bugs. Each release of user-interface software, even an "internal" release, should have a new

version number associated with it, and it should be
displayed prominently so that users can easily include it
in all bug reports. This is such a simple technique that it
is frequently overlooked entirely, much to the
developers' eventual chagrin when it ceases to be
possible to determine whether a bug is believed to be
fixed in the version that a particular user has on his or
her machine.

Source Control. It is amazing how many projects still have
multiple programmers overwriting one another's files
without ever realizing it. There are many software
packages, including some quite good ones that are in the
public domain, that help prevent this by forcing
programmers to "check out" and "lock" source files
before they may modify them. Such tools should be
considered as essential and as basic as compilers,
debuggers, and editors for the software development
process whenever more than one programmer is working
on the same program.

Coding Conventions. Although it seems trivial at first
glance, a set of agreed-upon conventions for a project's
source code can yield significant benefits in readability,
consistency, and maintainability. Coding conventions
typically specify such things as how variables are
named, how modules are named, where (i.e., in what
files) declarations and modules are found, and what kind
of internal and external documentation is required.
Adhering to such conventions is nearly painless if the
conventions are well specified in advance, and
impossible otherwise. Such conventions pay off
substantially when it comes to reading the code of other
programmers in the project. It doesn't much matter
which conventions are used; as long as the entire project
is using the same conventions, each programmer will
have a much easier time reading the other programmers'
code.

Code Reviews. The process of having another programmer
look at one's code, either formally or informally, is
always educational and beneficial. Even more beneficial
is the knowledge, when writing a program, that someone
else will actually be reading it soon.

Specifications. The level of formality may vary greatly, but
some kind of specification in advance of writing code is

essential in any large project, yet is routinely omitted in user-interface development. We have already noted that user interfaces cannot really be well specified in advance, but this is hardly an excuse for failing, as some programmers do, even to outline one's code in advance of writing it.

Egoless Programming. This term, coined by Weinberg (1971), may seem like a contradiction in terms to anyone who has ever met a programmer, but the idea is sound. The notion is that one's ego should not be heavily invested in any particular piece of code, and that one should always be eager to hear suggestions on how the code could be improved or could have been written better. Although utopian in the extreme, it is a worthy ideal and one worth striving toward.

Tools

Although standards and methodologies may be mixed blessings for user-interface designers, good tools, particularly software tools, are indispensible. Since the essence of user-interface development is rapid prototyping with frequent revision, any tools that help to speed up this process are worth their weight in gold-plated bits.

There are, in fact, a wide variety of tools aimed at the user-interface developer. Some of these tools are positively awe-inspiring in their level of sophistication, apparently rendering the construction of a user interface a nearly trivial task. Unfortunately, this appearance is deceptive. Most tools that claim to simplify user-interface development or facilitate rapid prototyping are worthless, victims of a few basic and fundamental flaws.

The most common flaw is *incompatibility*. Imagine, for example, that your job is to develop a prototype user interface that runs in a particular computing environment, on a particular vendor's hardware. Even if your interface is clearly understood to be a *prototype*, it should be built, if at all possible, on the target system. Building it on some other hardware or in some other software environment will lead to subtle or not-so-subtle differences, and even very subtle differences can substantially alter the way users perceive your software. Most of the fancy user-interface development aids on the market, you will find, only run in relatively restricted environments, and are thus rendered immediately useless for a wide range of projects.

A second common flaw is *constraint*. Often, a user-interface development tool makes it incredibly easy to develop an application, so long as you don't mind if it looks almost exactly like all the other applications developed with that tool. This may be fine for some projects, but is generally unacceptable to the designer when the user interface is intended to be a significant "value added" aspect of the final product. Good user interfaces often have their own unique style, or "signature." Moreover, different general styles of interface may be preferred for different kinds of applications. Tools that constrain your interface to a specific style or appearance aren't really prototyping tools at all, because you can't apply any lessons learned from your "prototype" to the next version of the system. Instead, they are best understood as tools for building simple interfaces quickly, when the quality, look, and feel of the interface are not terribly important to the final product. They are, in a sense, automated user-interface designers, but, fortunately for the humans who make their living building user interfaces, they aren't very good at it. They are essentially useless for the serious designer of user interfaces.

A third and perhaps most deadly flaw is *sloth*. It is commonly observed that the speed of a program goes down as its level of abstraction and generality goes up. Some user-interface prototyping tools extract a fearful price in performance. This may be acceptable if the prototyping can be done on a significantly faster machine than the ultimate target machine, but will probably be unacceptable otherwise. Users are extremely sensitive to performance issues, and it is difficult to get a clear understanding of the basic quality of an interface if it is prone to long and maddening delays in basic operations. Rapid prototyping is valuable enough to be worth paying for, so some performance penalty is not unreasonable. However, it is vital to be able to notice when the performance becomes so bad that users are responding to that problem rather than to problems with the interface itself.

Even if these ambitious tools don't live up to all their hyped-up billing, more basic tools are relied upon by every programmer. Any programmer developing a large user-interface system without the benefit of such tools as compilers for high-level languages, text editors, debuggers, icon editors, user-interface toolkits, and program consistency checkers, is programming in the dark. Any programmer who isn't at least aware of the potential benefits of cross-referencing programs and structure editors is ignorant of his own profession. Such tools should be simply taken for granted; one cannot program as productively without them, and if they are available, one shouldn't even try.

User Studies and Experiments

Although its role can easily be exaggerated, there is much to be said for careful and controlled observation of the users of a software system. A balanced understanding of what can and cannot be accomplished via user studies is essential to exploiting the opportunities they present.

Before we start, we should recall why user studies don't solve everything. First, they rarely provide *any* specific answers to important questions. On the few occasions when they provide really unambiguous results, they are often answering questions so narrowly focused as to be seriously limited in their applicability. Moreover, the human-computer interaction literature is filled with studies that exaggerate or overgeneralize their findings, so the reader must consider each claim quite skeptically.[1]   Good, conservative user studies nearly always produce more questions than answers.

---

[1]Why are there so many suspicious studies reported in the literature? There are several reasons. Desperate people, such as those facing thesis or tenure committees, can easily stretch the truth without admitting even to themselves that they are doing so. Most studies are considered "publishable" if they show a probablity of greater than ninety-five percent that they have found a statistically significant effect; this means that even if no corners are cut and everything has been done "perfectly," one in twenty studies may well be wrong. Unscrupulous researchers, aware of these odds, may further stretch the truth in the knowledge that such statistical claims aren't likely to count against them in any court. Still, most of the problems with experimental studies probably aren't caused intentionally. More often, researchers can corrupt their data by accident, or misrepresent their data due to a misunderstanding. In short: five percent of the experiments should be expected to be wrong on a purely statistical basis. Beyond this, some researchers are desperate enough to fake their results, more are careless enough to ruin their data, more still are ignorant enough to misinterpret their data, and nearly everyone is fallible enough to misinterpret others' data. The net result is that you should view all unreplicated published reports of user experiments with a decidedly skeptical eye. On the other hand, don't assume it's *all* worthless: a few studies have been reliably and repeatedly reproduced, and offer genuine guidance to the designer.

Why, then, are user studies worthwhile? The simplest answer is that they are useful precisely *because* they produce more questions than answers. In the best cases, they focus our attention on the important problems we might otherwise overlook. Often they help to eliminate incorrect hypotheses about the system being studied. Frequently, the very act of careful observation yields serendipitous insights that were never sought but are tremendously useful.

Controlled Experiments

Controlled experiments may be defined as those that are planned in advance, with a few well-defined specific questions to be answered. Typically, they take place in an artificial setting, or laboratory, with moderately artificial tasks, and produce data in the form of detailed timings, error rates, or some other numerical and (in principle) repeatable measurement. The goal of such studies, in general, is to obtain statistically significant results—that is, results that offer a high degree of confidence that the researcher has detected a meaningful effect.

The hardest part of a controlled experiment, it turns out, is its design; a poorly designed experiment will almost never yield meaningful and reliable data. One must understand clearly, from the outset, precisely what one wants to learn. Typically, one investigates a hypothesis such as "if the user has more X, his Y value will be higher." That is, the experiment will vary one aspect of an otherwise controlled setting, and will seek to prove that this variation has an effect on another particular aspect of the experiment.

The great fly in the ointment of controlled experiments is always *variation*. One might use a controlled experiment to "prove" the relationship of gender and typing speed, for example, showing that women type much faster than men. Yet if the women in the experiment were all professional typists, while the men were not, this would clearly be an independent source of variation in typing ability. The experiment would really only have proved that female professional typists can type faster than male nonprofessionals—but it would be easy to present the results in such a way as to make it look as if gender were the only relevant variable.

Variation may indeed occur due to the deliberate and planned manipulation of the experiment; this is the kind of variation that experiments seek to measure. Variation may also be due to "covariance"—joint effects of multiple variables, any of which may or may not have been a controlled factor in the experiments. Individual differences, such as the differences between typists and nontypists, may also account for variation in experimental results. Finally, experimental error, bias, and pure randomness are a source of a great deal of the measured variation in many experiments.

Variation is always possible, and there is no formula for avoiding it. Those who conduct experiments and those who read about them should therefore always be alert for every possible alternative explanation to the "obvious" explanation the experimental results offer.

Error, bias, and unexplained variance may creep into experiments in any number of ways. There is the infamous Hawthorne effect, which says that people tend to respond positively to *any* variation if they are led to expect improvement. There are classification errors, in which experimenters incorrectly judge the category into which a subject should be placed. (Even as basic a matter as gender can be misjudged by an experimenter, since most of us are too polite to inquire about an androgynous subject's gender. Most categories are considerably *more* prone to classification errors than this one, however.) Experimenter bias is always a possibility; the experimenter may have noticed fewer mistakes in the typing of those who seemed like professional typists. Errors are introduced simply by the sampling process, because even the best typist will have a slow day occasionally. Errors are sometimes introduced by the ordering of the subjects, because the experimenters may grow either better or worse at collecting the result as they grow more experienced and more fatigued. Measurement errors, caused either by defective apparatus or by overestimation of one's measurement precision, are another possibility—even stopwatches are known to have introduced significant errors into otherwise valid experiments. Experiments are so easy to misrepresent and misinterpret that we cannot even begin to list all the posibilities here. A marvelous introduction to the skeptical interpretation of experimental results is provided in a delightful and readable book by Schuyler W. Huck and Howard M. Sandler (1979).

In the end, controlled experiments can, with a lot of work, provide some fairly definitive answers to broad questions, though not always the same questions being asked at the outset of the experiments. The well-known studies by Teresa Roberts and Thomas Moran (Roberts 1979; Roberts and Moran 1983; Borenstein 1985b), which sought to provide an empirical basis for comparison among text editors, never managed to answer such broad questions as "which editor is best," but did provide persuasive evidence that the addition of a mouse is a signficant improvement to nearly any text editor. Similarly, my own studies of online help systems (Borenstein 1985a) never managed to prove that one help system was better than the others, but provided strong evidence that the quality of the actual help texts was far more important than the mechanisms by which the texts are accessed.

Perhaps the real value of experiments, however, can be found not in the answers they provide but in the new perspective they can offer on familiar problems. If one carefully reads the accounts of an experiment, and asks the hard questions that help ascertain whether one is going to believe the results, then the value of this reflective evaluation, and of the time spent thinking about how to answer the question, may be of great value in itself. Thus it can be educational and enlightening to read about even the most inconclusive or flawed experiments, provided that one thinks about them critically, yet with an open mind.

Passive Observation

Whereas controlled experiments focus a great deal of attention and effort on answering specific questions about user behavior in a carefully specified setting, the simpler alternative of *passive observation* is often at least as useful, if not more so. The goal of passive observation is usually to answer questions about the overall quality and utility of an existing system for its user community. Typically, such observations are conducted by instrumenting an existing system or by simply "looking over the shoulders" of users as they go about their normal use of the system.

Passive observation has the benefit of being simple, generally nonintrusive, and natural—the tasks observed are the ones that are actually useful, rather than devised solely for an experiment. As in controlled experiments, passive observation can yield highly detailed and specific numeric data—for example, about how often and how successfully certain features are used.

Such observation is extremely common, in large part because it seems so simple that anyone can do it. Nonetheless, there are a few pitfalls. First, there are ethical problems: this kind of observation should never be made without the subjects' consent, and should always be sensitive to privacy concerns. Additionally, it is often easy to miss the most crucial data in such observations. On the other hand, it is easy to collect so much data from instrumented systems that one can only be overwhelmed by it all, and unable to assimilate it. Another potential problem with passive observation is *intrusiveness*—the experimenter must take care that the mechanisms by which subjects are observed do not themselves interfere with the subjects' performance, either by virtue of their visibility or by the extent to which they slow down or alter the underlying system being studied. Finally, but often most important, it is sometimes extremely difficult or impossible to alter the system being studied in such a way that the data can be collected at all.

For passive observation to be successful, it is extremely important to select carefully the data being captured. The observer must begin with a set of clear questions, problems, or hypotheses regarding the system being studied, and must understand exactly how the data relate to these hypotheses. Even then, it is hard to be certain that one's data do not suffer from a "Heisenberg effect," in which the mere act of measuring something fundamentally alters the thing being measured. (For example, logging events by sending mail—a relatively costly way to do logging—can slow a system down so much as to make it unusable.)

Despite all these problems, passive observation studies are still well worth pursuing in many contexts. They are invaluable in developing an understanding of how deployed systems are actually used in the field. They may also be used, for example, to classify users by expertise and work patterns, or to characterize the extent and purpose of system use in a given organization.

An excellent example of this kind of study is provided by Draper's report on the nature of expertise in the UNIX operating system (Draper 1984). In this experiment, researchers sought to develop a detailed picture of those people who are commonly referred to as UNIX "wizards"—that is, the real experts. They hypothesized that the wizards would share a common core of expertise, and that this core could be characterized as the set of things one needs to know in order to be a UNIX wizard. In order to figure out what was in this core, they instrumented an entire UNIX machine to capture information about all of the commands that the wizards used. They then tried to find the common overlapping core of UNIX knowledge that the wizards all shared. Much to their surpise, however, the results showed that the core was vanishingly small. The expertise of the wizards scarcely overlapped at all!

·The core set of knowledge shared by all the experts was, in fact, so small as to be easily mastered by anyone, and in no way sufficient to define expertise in UNIX. In this case, because the experimenters had begun with a clearly defined hypothesis and plan, they had been able to collect precisely the data that would be useful, and thus used passive observation to obtain a valuable and surprising insight.[2]

Detailed Observation

Another common kind of user study, which may be referred to as *detailed observation*, takes another approach entirely. Instead of obtaining the kind of numerical data that is available through the previously described techniques, detailed observation seeks to gain deeper insight through intensive study of a single user. Typically, such studies include a large number of questions about an individual's use of a system, and often are accompanied by monitoring the user's progress from the time he or she begins to learn the system through his or her maturation as an expert.

Such studies are, of course, extremely intrusive; the subject is inevitably aware of being under the experimenter's microscope. The tasks in such an experiment may be more-or-less natural, depending on whether the experimenter is determined to see the subject's behavior using features he or she might otherwise never try. In such experiments, careful measurements are possible, but not terribly useful, as it is almost never possible to conduct such experiments on a large-enough sample of users to obtain statistically significant results. Instead of hard data, the goal of such observations is usually a deeper

---

[2]Follow-up studies and interviews convinced the researchers that the idea of a "core" was fundamentally wrong, and that the best definition of a UNIX wizard was someone who knew how to find out how to do something on UNIX using the online or paper manual. In other words, expertise was best defined in terms not of what the expert already knew, but of how easily the expert could find things out.

understanding of how users cope with the system. The most valuable outcome of these observations is often in the form of comments, anecdotes, and sudden insights. Unlike the kinds of experiments discussed previously, detailed observations require relatively little effort to design. They are more typically "fishing expeditions," where the experimenter has relatively few preconceptions about what will be learned.

One popular technique in these observations is the "speak-aloud" protocol, where users are told to work "normally" except that they should keep up a running verbal description of everything they are doing and thinking. Such accounts, when taped and transcribed, can yield considerable insight into why users make the errors they make, and what they're looking for when they're "stuck" in the middle of a task. Obviously, this technique can hardly be considered "natural," and may alter the user's behavior substantially.

Among the pitfalls of this approach is the aforementioned "Heisenberg" problem of experimental interference with the user's task—often the experimenter can change what happens merely by smiling! Furthermore, reliance on artificial tasks for such observations increases the likelihood of artificial insights—that is, insights that do not transfer to more realistic tasks. Subjects who feel as though they are being scrutinized under a microscope are prone to nervousness, which can seriously affect their performance. Finally, given the inevitably small number of data points in such observations, it is easy for the observer to fall victim to overgeneralization. One should remember that there are always users whose problems and opinions are highly idiosyncratic.

Although statistical significance is almost unheard of in such observations, numbers can still heighten the impact the observations have on the developers. If three of the four subjects observed had trouble with a given feature, that feature should certainly be examined closely. (If only one subject has trouble, the feature should still be examined, but the larger numbers do add weight to the case for devoting the time to further analysis.)

Detailed observation has been used to great advantage in studies of novices learning to use new interfaces. They have been used, somewhat more sporadically, to gain insights into how experts do their jobs. In the best of cases, a series of detailed observations can be used both to promote the evolution of a better interface and to reflect and document such improvements in later versions (Boyarski, Haas, and Borenstein 1988).

User Surveys

One of the most common and useful forms of user studies is the user survey. A survey simply asks the users a set of questions about their perceptions and use of the system under study. Such studies typically consist simply of having people fill out a survey form, and analyzing the resulting data. User surveys are an unobtrusive way to gather a good deal of subjective and nontask-oriented data about a user interface.

Within a survey, there are two types of questions that can be asked. Very specific questions—for example, in a multiple-choice format—are highly amenable to statistical analysis and the detection of trends. More open-ended questions yield less hard data, but often open the door to significant new insights.

As it turns out, there is a very large existing literature on the subject of survey design. The order and wording of questions turns out to be critical, and there are many other pitfalls awaiting the unwary user-interface designer who wants to use surveys as a tool to study his user community. Sophisticated surveys often employ prescreening of subjects, or sort the subjects on the basis of background information that is collected with the survey.

The results of user surveys may be skewed by variations in individual experience, skills, and temperament, by the environment or circumstances in which the survey was taken, and by subtleties in the way the questions are worded. The overall pool of subjects may be further biased by the method with which the survey is distributed—as a gross example, consider a survey about how, when, and why people use electronic mail: if such a survey were itself to be distributed via electronic mail, would this bias the results toward underrepresenting people who never or rarely use electronic mail?

In short, the number of potential problem areas in user surveys is so great that it almost always pays to consult an expert, both before the survey is distributed (so that the expert can look for flaws in the survey itself and in the distribution plan) and after the results are collected (so that the expert can help ensure a valid analysis of the data).

The great value of user surveys is the extent to which they can often reveal the degree of *subjective* satisfaction in the user community—something that is generally overlooked by other forms of user studies. They are therefore particularly valuable from a long-term planning perspective and a marketing perspective, since subjective judgments will tend, in the long run, to be the ones that matter most to the user community and to the success of software in the marketplace.

The truth is that software that scores well in user-studies experiments is not necessarily software that will be well-loved in practice. While more quantitative studies may be invaluable in improving a system and detecting its major flaws, it will never capture the subjective element that is so important. For example, in my previously mentioned dissertation work, controlled experiments were used to contrast a variety of online help systems. Surprisingly, a number of the systems that had radically different interfaces scored equally well on all the "objective" measurements. Nonetheless, follow-up user surveys revealed strong user preferences for the flashier, more highly developed user interfaces. Although the experiments showed that the fancy interfaces had little, if any, effect on actual user performance, the subjective surveys revealed that they would have had a significant edge in the marketplace because people *enjoyed* using them, whether they were really useful or not. This is precisely the kind of data that user surveys are good at providing.

Techniques

As time goes on, every programmer gradually accumulates a collection of techniques that can be applied repeatedly to a wide range of situations. User-interface programming is no different in this regard; there are established techniques, or "tricks," that can be used over and over again.

For example, it is well known that when some kind of time-consuming processing is being done, during which the user has to wait, some sort of visual indication of that fact should be presented to the user. Many graphical interfaces use a designated "wait" cursor, such as an hourglass, to indicate this condition. Less common, but extremely useful, is to provide a diversionary activity that does not compete for whatever resource is causing the delay. For example, if the delay is caused by waiting for a response over a network, or from some slow peripheral device, it is perfectly plausible to display some animated graphics on the screen, as animations typically make heavy demands on the CPU and display hardware but not on the peripherals. Similarly, if the CPU is heavily loaded but a relatively smart display is not doing much, it can be made to display an interesting picture while the processing is going on. The potential for such diversions would be far greater if hardware and operating systems were designed with them in mind. For example, a cheap secondary CPU could allow whole video games to be played while the main CPU was busy, and good operating system support for software interrupts would permit much more flexibility in starting and stopping diversionary activities.

Other established techniques for user-interface design include visual repetitiveness, to suggest analogies visually among separate pieces of software, and even such basic notions as always asking for confirmation before performing any destructive operations. Unfortunately, no definitive list of such standard techniques has ever been compiled, although Paul Heckel (1982) lists a substantial number of them.

Professional Communications

Perhaps the best aid available to the user-interface designer is close communication with other interface designers. Increasingly, this kind of communication, for various specializations within the computer world, takes place via electronic mail and bulletin board networks. Although taking part in such services is often regarded by outsiders as a waste of time, the specialized discussions can be extremely educational, and enlightened managers should encourage a moderate degree of participation in such electronic conferences.

Technical conferences, lectures, and informal meetings are also vital in forming professional contacts. A network of such contacts is useful not merely for politics (e.g., getting another job) but, more importantly, for developing a set of colleagues who can offer intelligent opinions and suggestions on difficult design decisions.

Of course, professional journals, such as those put out by organizations such as ACM and IEEE, and technical conferences such as the annual CHI conference, are another vital form of contact with one's fellow professionals.

Interesting Stories

As discussed in chapter 4, ''war stories'' are both favorites of programmers everywhere and valuable sources of practical wisdom and education. Unfortunately, there is absolutely no established forum for the interchange of such stories except private conversation, usually over a few beers. It is intended that the stories in this book will serve as a small step in that direction. A few other books, notably those of Tracy Kidder (1981) and David E. Lundstrom (1987) offer entertaining and useful collections of enlightening anecdotes.

To delete a file on the UNIX system, one uses the "rm" command. The "rm foo bar" command deletes the files named foo and bar, and "rm *" deletes all files in the current directory.

A few commonly used options complicate the syntax of this command. In particular, the "-i" option means that rm will ask for user confirmation before deleting files. However, the "-f" option will override "-i" and any other mechanism that would require user confirmation, and will cause the files to be deleted without question. Thus "rm -i *" will cause the user to be asked individually, regarding each file in a directory, whether that file should be deleted. On the other hand, "rm -f *" will delete all files in the directory without asking for any confirmation.

Warren Carithers told me about one of the most amazing rm-induced disasters, which has been fixed in some later verisons of UNIX:

> A student went to his instructor for help in removing a file named "-f" from his account. The instructor first attempted "rm -f," which didn't complain but also didn't remove the file. After a few similar attempts, the instructor fell back on the tried-and-true method of "rm -i *." Some time passed during which no messages appeared on the terminal; as the instructor began to grow uneasy, the next shell prompt appeared. A directory listing showed that only one file remained, named "-f." At this point, the student (who had been watching the proceedings over the instructor's shoulder) commented, "If you weren't my teacher, I'd think you just deleted all my files." The problem . . . was the interpretation by "rm" of the first file name, "-f," as an argument. The result was that the "-i" option actually given by the instructor was overridden by the name of the first file to be removed.

```
$ ls
-f      file1   file2
$ rm -i *
$ ls
-f
$
```

Chapter 20

The Ivory Tower

In order to become a doctor and help the sick of the ghettos, you must first study "irrelevant" subjects like comparative anatomy and organic chemistry. In the same way, you must give yourself up to the seeming irrelevancies of the past in order to be relevant to the present.

—Robert Brustein

Education has this sovereign purpose: to prepare one for more education. All else is subsidiary to this. Education should create hungers—spiritual, moral, and aesthetic hungers for value.

—Israel Knox

In industry, alas, it has long been taken for granted that the universities provide little practical training for programmers. Programmers often regard their formal training, if they have any, as little more than a bad joke. Statements such as "I learned more in four months on the job than in four years of college" are so common as to lead almost inescapably to the conclusion that something is seriously wrong with the way programmers are educated in our universities today.

The principal problem is that the skills taught in university computing programs are astonishingly irrelevant to the tasks of programmers in the real world. Of course, actual programming courses, in which students learn how to use one or more programming languages, are useful to those who want to program for a career. But beyond the mechanics of programming, the typical undergraduate curriculum has very little of relevance to the practicing programmer.

Consider the typical courses one must endure to obtain a degree in computer science. Beyond introductory programming, one typically immediately encounters a class in formal methods. Such classes go under different names at different universities, such as "Fundamental Methods of Computer Science" or "Foundations of Computing," but their content is fairly standard. They teach the student about data structures, abstract data types, analysis of algorithms, program verification and synthesis, and the mathematical theory of computation.

With the possible exception of data structures, which are essential to any serious programmer, these areas offer little that will be of any use to the practical programmer. While it is fascinating, from a mathematical viewpoint, to be able to characterize the intrinsic computational complexity of a problem, it is considerably more rewarding, for the programmer, simply to solve it in the best possible manner—where "best" often refers at least as much to the speed with which the solution can be implemented as to the ultimate performance of the program. The theory of computation characterizes only one of the relevant factors—the performance of the program itself—and even there, rarely offers anything of use because the programmer is not usually working in the uncharted waters where theoretical questions arise. The programmer working on a new spreadsheet or word processor is unlikely to encounter any questions of mathematical theory while trying to decide how best to present information to the user, for example. Theoretical considerations will be useful in certain important subproblems—optimizing screen redisplay, for example, or correcting spelling—but the efficient programmer can generally find the "right" answers to such problems in reference books, notably Donald E. Knuth's *The Art of Computer Programming*. The really useful skill, in such cases, is the ability to find the right algorithm in a reference book on computer algorithms, and this is certainly a skill that young programmers should be taught in school. The simple fact is that most practical programming does not involve much mathematical theory at all, and programmers can learn the relevant bits of mathematics the way engineers often do—with a relatively superficial understanding, but with a clear knowledge of how to find the answers in reference works.

Similarly, programmers are virtually never called upon to prove anything about the formal capabilities of their system, though the undergraduate curricula at some universities could leave one with the impression that this is their primary duty. While programmers would certainly like, for example, to be able to prove mathematically that their programs are bug-free, experience has shown this to be impossible in any practical sense. Real-world programs are both too large and too ill-specified ever to be proven "correct." Advocates of the formal verification of programs have long admitted that no program of any significant length has ever been proven correct, and they have even been known to state that this is because none of them *are* correct. They are probably right. Nearly every useful program has bugs, and nearly every program simple enough to be bug-free is likely to be of very little practical utility. In general, this doesn't matter, because big programs with occasional bugs have proven to be so useful in the real world. (Of course, there are times when this situation is less tolerable, which is why so many computer scientists were horrified by the "Strategic Defense Initiative" proposed in the 1980s, which would have placed history's most complex programs in charge of weapon systems that could destroy the planet.) Programmers don't prove programs bug-free because they don't write bug-free programs. Not only *don't* they write them, but they generally *can't* write them, or even describe how their programs would behave if they *were* bug-free. This is simply the nature of the programming art.

The universities deplore this fact, and teach students program verification as if the magic of that teaching will somehow transform the nature of the computing world, so that program verification will become an important part of programmers' lives. Instead, verification becomes a shared joke among programmers—one of many things they had to learn in school, but could then ignore. Teaching undergraduates about program verification makes about as much sense as teaching impoverished African children how to make money on the stock market. One can imagine some farfetched circumstances in which the skills might be useful, but the mere teaching of the skills does nothing to help create the circumstances in which they might be used. Knowing about the stock market is useless to people with no money, and knowing how to prove programs correct is useless to programmers in a world where programs are not correct and don't really need to be.

Historically, it is easy to see how we ended up with a computer science curriculum that teaches very little of practical value. Computer science has its origins in two very distinct academic disciplines, mathematics and electrical engineering.   The age of computing was born when such brilliant mathematicians as Alan Turing and John von Neumann had a series of insights into the nature of problems that can be characterized formally and solved via formal processes. Turing, in particular, proved that any machine of sufficient computational power was "universal" in the sense that it could answer any question that any other such machine could answer. The simplest "universal" machine, it turned out, was so mind-bogglingly simple that, in recent years, one has actually been built out of Tinkertoys for the Computer Museum in Boston. The computing era really began when mathematicians formally proved that relatively simple machinery could solve any computable problem.

At this point, practical issues began to take hold.   These theoretical "universal" machines, it turns out, have infinite time and memory in which to perform their tasks. In the real world, the value of a computing machine is directly related to the size of its memory and the speed of its operations. Improving the practical capabilities of real machines quickly became the province of electrical engineers, who brought with them their own methods and attitudes. By and large, the engineers were very practical people, eschewing formalisms in favor of whatever worked well and quickly. But though their methods worked well for computer hardware, they were less successful with complex software, which became the domain of the mathematicians, largely by default.

One cannot really fault a mathematician for seeing a computer program as a mathematical object. After all, mathematicians see everything as mathematical objects, and rightly so.   When mathematicians rigorously consider a program, they may achieve all sorts of insights about the nature of programs, the nature of the problem addressed by the program, and the nature of computation. This is all good, in the same way that any mathematical progress is likely to be a good thing, an augmentation of human knowledge. However, none of this implies that a mathematical perspective is necessarily the most useful one for the practicing programmer. The derision with which many professional programmers view the formal methods in which they were trained is a strong indication that the training was inappropriate and misdirected.

Indeed, even the academic computer science community seems to know that something is wrong. Dissatisfaction with the way that computer science is taught is widespread. In a recent and widely read article, the eminient computer scientist Edsger Dijkstra rails against what he calls "the cruelty of teaching computer science" (Dijkstra 1989). But the solution offered by Dijkstra is, alas, the only one that is ever likely to be offered by those overly steeped in the mathematical approach to computing: an *increased* reliance on formal methods and proofs. To the practicing programmer, the situation would be laughable if it weren't so distressing.

The situation has been tolerated, in large part, through lack of a better alternative. The students emerging from a typical undergraduate program in computer science or software engineering may not have learned the most useful skills for their chosen career, but they have at least proven themselves capable of mastering highly technical material. Indeed, the role of the universities might best be viewed as one of prescreening candidates for future employment: the university graduates may not know what they need to know, and may not be guaranteed to succeed in the workplace, but at least they're not as bad as some of the people who couldn't make it through the university at all. (This may even be a charitable view of the university's role, since it is unclear that failure in a university computer science program is a good predictor of failure in real programming jobs: many great young hackers are actually dropouts from undergraduate computer science programs that they see as infuriatingly irrelevant to their chosen careers.)

The interesting question, then, is whether the universities might actually be doing something more useful. Is it possible to train young men and women in skills that will actually increase their efficiency as computer scientists? To answer this question, we must either invent a new discipline of computing out of thin air or seek a model from another discipline. The existing paradigms for computer science are derived primarily from mathematics, with a bit of electrical engineering thrown in. Are these really the closest analogues to what programmers do, or are they merely a historical starting point?

To ask the question is to begin to answer it. The university is, in fact, filled with alternative teaching paradigms that have much to offer computer science. Architecture, for example, shares much with programming. Architects must design a building in accordance with the laws of physics and the realities of economics, lest it collapse when built or be too expensive to build in the first place. Yet if they are trained only in such "formal" methods, they will build only ugly boxes that will blight the landscape. Instead, architects are also trained in aesthetics and the history of architecture, and are encouraged to develop their own creative impulses. The architecture curriculum recognizes the need to encompass both science and art. Certainly it has lessons to offer the programmer who must build user interfaces that are cost-effective, reliable, usable, and well-liked.

If the computer science community could get its head out of the sand long enough to consider its place in the academy seriously, it would find numerous other role models among the arts and sciences. Musicians, for example, must be trained in music theory, which is essentially a mathematical discipline, but they must also be coached to practice endlessly and constructively. The typical music student practices daily, and meets at least weekly with a "master" who can offer useful tips and feedback. Would not such a regime be equally beneficial to young programmers trying to master their craft? Similar lessons can be had from anthropology, art, drama, and probably nearly any other department in the university.

Fundamentally, today's computer science departments, which are extremely young by university standards, suffer from the historical accident of their mathematical origin and from a lack of time to reflect on their inherited paradigms. Eventually, it seems inevitable, computer science will develop its own methods and models for training both programmers and researchers. To date, however, it has failed to address seriously the difficult questions of training practical programmers. In the face of evidence of its failures, the first response of some of its luminaries is simply to intensify the old reliance on formal methods. But it is unlikely that the problem will ever be solved by mathematicians alone, as it is unlikely that mathematics alone provides the right perspective for programmers. Only by a careful consideration of a wide range of disciplines is it likely that computing will be able to find its own role as an interdisciplinary area of study within academia.

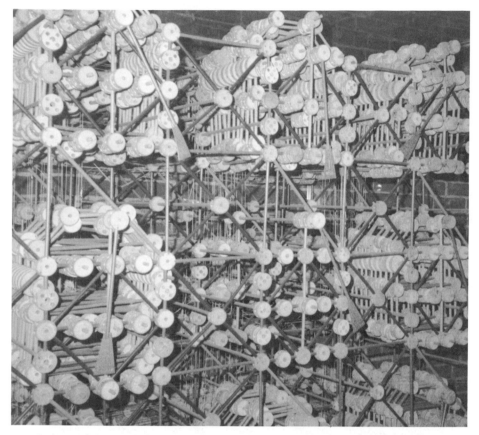

In one of the world's most memorable attempts to demystify computing, Danny Hillis built a "computer" out of Tinkertoys, with a hard-wired program for playing tic-tac-toe. This technological marvel, on display at Boston's Computer Museum, also unintentionally displays the gulf between what an engineer calls "success" and a user's definition of the term. The sign on the display reveals the sad truth:

> *The machine never lost, but played unreliably*
> *because the fishing line tended to loosen and slip.*

In other words, the programmer built something truly wonderful, but you can't actually use it to do what you want to do. Sound familiar?

Nine subscribed folders with new messages	
☒? CONSOLE (Ask-Subscribed, No New) ☒? TODO (Ask-Subscribed, No New) ☒✓ mail (Personal mail, 1 new of 1182) ☒✓ org.postman.bbmaint (Has New Messages) ☒✓ org.itc (Has New Messages)	Punt!

✓ 1-Dec-88 *Re: more on 7.0* - Nathaniel Borenstein (318+0)
✓ 1-Dec-88 *FYI -- ThisDomain* - Uncertain Borenstein@and (199+0)

X-Andrew-Authenticated-As: 327;andrew.cmu.edu;User 327 in cell andrew.cmu.edu
Return-path: <nsb+@andrew.cmu.edu>
X-Andrew-Authenticated-as: 327;andrew.cmu.edu;Nathaniel Borenstein

Electronic mail systems are, so far at least, infamous for producing long and inscrutable return addresses that make it hard for mail recipients to figure out who their mail is actually from. The Andrew Message System devoted a substantial amount of code to trying to extract an actual human-readable name from such headers, with generally good results. However, the methods used were, of necessity, ad hoc, as it was not possible to find the correct human-readable name all the time. Sometimes the attempt backfired with amusing results such as the one shown above. In an even sillier incident, the system shortened the ugly address

    POSTMASTER <ROOT@ELEPHANT-BUTTE.SCRC.SYMBOLICS.COM>

to the delightful shorthand

    POSTMASTER@ELEPHANT-BUTT

This sounded so unlikely that the recipient simply assumed he was the victim of a practical joke, rather than a software error.

Chapter 21

People Are Perverse:
Designing for the Fickle User

Practice gives duration to problems. Practice is total inattention. Never practice: you can only practice mistakes. Learning is always new.

—Krishnamurti

I do hate sums. There is no greater mistake than to call arithmetic an exact science. There are permutations and aberrations discernible to minds entirely noble like mine; subtle variations which ordinary accountants fail to discover; hidden laws of number which it requires a mind like mine to perceive. For instance, if you add a sum from the bottom up, and then again from the top down, the result is always different.

—Mrs. La Touche (nineteenth century)

User-interface design is, at bottom, a very tricky business. There are as many techniques for achieving a good design as there are good designers. It was never my intent, in writing this book, to put forward a definitive solution to the innumerable problems of interface design. This book was motivated by a more modest and proximate goal, the debunking of a number of myths and illusions about the process by which good interfaces are and should be designed.

The wealth of negative prescriptions in this book, it is hoped, may help a few readers to avoid some nasty pitfalls in user-interface design—pitfalls that I discovered headfirst in many cases. Out of all these warnings, counterexamples, and horror stories, however, a certain amount of positive advice can also emerge. The challenge, to a suitably humbled designer, is to design intelligently a user interface in full knowledge of the myriad ways in which he or she is fundamentally unprepared to do so.

Although user-interface programming seems unlikely to become a clear and rational process anytime soon, this does not prevent the occasional success story. What is needed, if designers are to increase their chances of success, is an "algorithm" for productively thrashing about in the dark until they stumble on a solution. Such an algorithm might be referred to as a methodology for human-oriented software engineering.

Toward a Methodology for Human-Oriented Software Engineering

The first thing that must be recognized is that using the kind of terminology that is familiar to engineers does not in itself turn any arbitrary subject matter into an engineering discipline. As much as engineers might long for a cleanly specified methodology for building user interfaces, the best they are likely to get is a few guidelines or a model that points them in the right direction. In the end, we're less likely to "engineer" user interfaces than to engineer the process by which a team, including some people who are decidedly not engineers, build such interfaces.

Another way of putting this is that we want to encapsulate the irrational, or human, element of the design process, so that the rest of the process may be rationalized. As engineers, we are thus admitting that artists have a role in the design process, but are trying to limit the extent to which this artistic component prevents us from using a fully rational development methodology.

In this spirit, we will consider a bare-bones methodology for user-interface development. This methodology reflects ideas found in various methodologies that have been presented in the research literature, and is given here to serve as a starting point for those who would build user interfaces as a mixed enterprise of engineers, artists, and social scientists.

A Methodology for User-Interface Development

How can the unpredictable whims and responses of humans be factored into the process of developing and maintaining user-oriented software? In the absence of a completely satisfactory answer, an organization could do worse than to base its product development on the general plan outlined in the sections that follow.

Definition Phase

The first and most important phase of any software development project is the definition phase. It is at this point, before any code has yet been written, that the development team first attempts to describe just what it is trying to build. The initial descriptions are rarely coherent, but serve primarily to clarify the developers' thinking. Ideally, early attempts at definition will be circulated widely among potential users of the yet-imaginary system, in the hope of detecting, at this early stage, the greatest possible number of misconceptions about the nature and utility of the system being built.

The definition phase is best understood as a process of frequent iteration between *needs assessment*, in which the ultimate requirements of the system are considered; and *functionality specification*, in which the developers' current best understanding of the system requirements are set down in a form that is comprehensible to those whose advice is desired—notably potential users.

The definition phase can be defined algorithmically, in pseudocode, as follows:

while (*doubt-remains* **and** *time-remains*) **do**

    begin (*definition-phase*)

    1. Assess needs. Talk to potential users about their needs and wishes. Observe users of manual or competitive systems to see what they actually do, and compare this to what they *say* they do. Think about additional features that users have not asked for but might nonetheless appreciate.

2. Specify functionality. Avoid the trap of writing overly formal, engineering-style documents that nobody will want to read. Instead, write complete, concise, well-organized lists of things the program must do, will not do, and might or might not do. As the system evolves, this document will be kept up-to-date, slowly changing from a "wish list" into an exact specification of what was actually built. Circulate these specifications to your users and return to step 1.

end (*definition-phase*)

The meaning of the "while" loop, in this case, is that these two steps are to be repeated many times, until either no time remains for the specification phase or further iteration seems likely to be unproductive. In practice, the latter is very rarely the case; more often one declares the definition phase to be completed only because deadlines are pressing. The quality of the definition, therefore, is critically affected by the amount of time that was initially budgeted for completing it.

Prototyping Phase

At this point, it is crucial to keep in mind the distinction between a prototype and a product. A prototype is an experiment, an uncertain step in what one hopes is the direction of a product. As has been emphasized throughout this book, prototypes should be written quickly and considered disposable. The speed with which a development team can write, evaluate, and dispose of prototypes is one of the most critical factors determining both the ultimate quality of the interface and the timeliness with which it will be delivered. This is why tools for rapid prototyping are so important.

while (*time-remains* **and** *ideas-remain*) **do**

    begin (*prototyping-phase*)

    1. Design a user interface. At this point, the thing to do is make your best guess, based on the specifications and your own experiences, and write down some rough outline of what it is. Since this is just a prototype, nothing formal is really appropriate; you need only write down enough to be reasonably certain that your development team shares a basic common idea of what the interface will look like.

    2. Build a "quick-and-dirty" system. The prototype should be built with relatively little concern for robustness or completeness, although it must, of course, run well enough to permit users to assess its interface. Although almost any corners may be cut legitimately at this stage, it will behoove the developers to keep a complete and rigorous list of all known inadequacies in the prototype, just in case it doesn't get thrown away as planned.

    3. Let users play with the prototype. There is a natural tendency to want to "polish" the prototype a good deal before unleashing it on even the friendliest of users. After all, no one wants to subject others to bugs or expose one's own mistakes unnecessarily. However, this tendency must be resisted. The sooner the prototype gets into the users' hands, the more useful feedback the developers will get. It should also be noted that while this stage is easily important enough to justify paying the users for their time, it is somewhat dangerous to come to rely on a regular staff of in-house paid "guinea pigs" for such testing, as repeated testing will inevitably reduce the resemblance of such a testing community to the real targeted users. Pay special attention to the users' *first* reactions, because they can get used to anything, however bad it is, in a surprisingly short amount of time. If you have a professional training staff, ask them what aspects of the system are hardest to teach; these are probably also the aspects that are hardest to learn.

4.   *Observe the users.*   As unobtrusively as possible, the developers and testing staff (if there is one) should carefully observe the users interacting with the prototype. Special attention should be paid to the methods by which users accomplish goals. If these methods are not those intended by the designers, or if they involve substantial and repeated errors on the part of the users, the relevant part of the interface probably needs to be redesigned.

5. *Collect user feedback.*   Ask the users a series of questions, both specific and open-ended, about the prototype. Make sure to give opportunities for both in-person and anonymous feedback; different kinds of feedback from different people will often require different channels of communication.

6.   *Reconsider the specs.*   Compare what you learned from the users with the specification you previously developed (in the definition phase). Adjust the specification accordingly.

7. *Back to the drawing board.*   Reconsider the interface design in light of the experience with the users, and return to step 1.

end (*prototyping-phase*)

Like the definition phase, the prototyping phase of development is unlikely ever to terminate naturally. At some point, production deadlines or waning interest will require that the prototyping phase be declared complete. (Conceivably, one could devise a user-satisfaction index by which one could declare that the prototyping phase ends—for example, when ninety percent of the users consider themselves happy with the interface—but such measures are tricky and unreliable.)

If a definite amount of time is clearly budgeted for prototyping when the project begins, this will at least help to avoid arguments over whether enough prototyping has been done. The great advantage of simply placing a time limit on the prototyping phase is that it frees the developers from having to answer the unanswerable question, ''Are we done prototyping yet?''

Production Phase

When, for better or worse, the prototyping phase has been declared over, the next step is to build a product. At this point, the team's energies should be refocused. Instead of concentrating on improving the user interface—which will certainly still have room for improvement—the team should focus on creating a robust, stable, efficient, and reliable implementation of the best prototype that was produced.

Ideally, this will involve throwing away the prototype and rewriting it from scratch. Alternatively, however, it can begin with that prototype and its list of known flaws, which was collected during the prototype's development. That list can be used as an initial specification of the difference between the prototype and the desired product.

In the production phase, the developers should think and behave as software engineers rather than as artists. The target program should be considered to have been nearly fully specified in the form of the final prototype, so that the goal of this phase is the engineering realization of a completely specified target system.

Of course, that won't really happen. The system will continue to evolve, albeit much more slowly due to the convenient fiction that the evolutionary phase has stoppped. As the system slowly evolves, however, the specification should always be kept up-to-date with reality. As features are dropped, the specification should be annotated to explain *why* they were dropped, so that the same mistakes won't be repeated later. Similarly, when new features are added, the motivation should be explained.

Maintenance Phase

Once the product has been built, the project enters the last phase of its life cycle, a phase most programmers prefer not to think about—the *maintenance* phase.   Software maintenance is unpopular with programmers because it is thankless and often difficult. With each new change, a program often becomes more complex, more difficult to understand, and more susceptible to mysterious new bugs. Software engineers often neglect maintenance issues, sweeping them under the carpet by saying that if only programs were built right in the first place, they wouldn't need so much maintenance.

Another possibility, however, is that maintenance becomes problematic because of confusion between such obvious necessities as bug fixes and component upgrades—which might collectively be referred to as "necessary maintenance"—and more optional activities such as extensions and feature enhancements.   The necessary maintenance serves to fix problems that are essential to the functioning of the product.   The optional maintenance addresses problems that are left over from the specification and prototyping phase—in essence, they are back-door attempts to return the product to those early stages of its life cycle. It is not surprising if relatively fundamental and experimental changes cause serious problems of maintenance.

It helps therefore, to maintain some kind of distinction between appropriate and inappropriate maintenance activities, much as one cuts off experimentation in the transition from prototype to product. The disadvantage of this approach, of course, is that one accumulates an ever-growing list of fundamental problems with the system being maintained, none of which can be legitimately addressed under the rubric of "maintenance." This does not, however, mean that they can never be addressed at all. Rather, they are best addressed as part of the first phase of a whole new product cycle, which is geared toward building the successor of the system being maintained, or at least something that is recognized, developed, and supported as a *major* new release.

While (*economic-justification-remains*) **do**

    begin (*maintenance-phase*)

1. Collect suggestions. Establish channels whereby users can easily communicate bug reports and feature suggestions to the maintenance team. Once these reports are received, it is surprisingly important to make sure that they are well organized, preferably into some kind of simple list or database from which they can be easily sorted by topic, severity, and difficulty. If such a list is not kept, the maintenance team will probably end up fixing things in a very ad hoc manner, with no clear notion of priorities or long-term plans. (It is worth noting, however, that you shouldn't need to get technology-happy with an extremely elaborate database mechanism. If you can't manage these lists in a simple sortable file or database, your system may already be hopelessly incoherent, and you should consider starting over with a replacement product.)

2. Choose what to fix. First, the most important bugs and suggestions, which we will call the *major flaws*, should be isolated from the others. Then, these crucial bugs should be sorted to differentiate the fixes that can be made relatively easily from those that will require substantial or widespread modifications. We will call the former category *tractable major flaws* and the latter category *intractable major flaws*. Make sure to consult the specification to find out if there was a good reason why suggested improvements weren't a part of the design in the first place.

3. Is it time to start over? If the set of intractable major flaws is sufficiently large or critical, it is probably time to consider a major redesign of the system—either a major new version, or a replacement system. This prospect always sounds intimidating, but the only alternative is to try to address the intractable major flaws in the context of routine maintenance. This is what typically happens in large systems, and it often implies systematic changes with far-reaching and unpredictable consequences. The impact of these changes may be great enough that the effort involved is comparable to what would be involved in a major new version of the system, but the result is likely to be a far less clean and maintainable system than would result from a rewrite. If a major starting-over is called for, one team should be left to do maintenance on the old version while another team starts over with the definition phase for the new product. It is particularly worth noting that the complex nature of user-interface programming, on most systems, implies that nearly any substantial or paradigmatic user-interface flaw is best considered an intractable major flaw, to be deferred until the next major redesign of the system.

4. Perform upgrades. The tractable major flaws should be fixed. Once this is done, as many of the minor flaws should be fixed as time permits. The fact that time never permits all of the minor flaws to be fixed simply underscores the importance of a carefully maintained and prioritized list of system flaws. Of course, the specification should be updated whenever the software is modified.

5. Test the new version. This is harder than it sounds. The unfortunate truth is that every time programmers make a large number of improvements to a program, they introduce some bugs as well. (The ratio of bugs fixed to bugs introduced, though hard to measure, is a good index of the *maintainability* of the system.) Thus each new release requires "regression testing" to make sure that what worked in the previous release will continue to work in the new release.

6. Release the new version to the users. This promptly starts the cycle over again.

end (*maintenance-phase*)

Following procedures like the ones outlined in this chapter do not, of course, even begin to guarantee the success of a software development project. There are no guarantees. What they more modestly offer is a framework in which to do what user-interface developers have gradually discovered must be done, which is to try and try again until the software finally becomes sufficiently usable to satisfy the needs of the various communities of likely users. Because managers often feel the need for a framework or "methodology," the absence of a flexible methodology such as this one can lead to the adoption of a more rigorous and harmful one. In this sense, the greatest benefit of the methodology outlined here is that it doesn't do as much harm as others that are commonly advocated by more formally-minded software engineers.

To: cfe
Subject: beroot
CC:

| Will Keep Copy |
| Will Clear |
| Won't Hide |
| Reset |

Apparently, some freshman has been using your beroot. I suggest you put some security on the thing which isn't trivial to walk around. If you want, I

Unrecognized validation code 810973798; please confirm:

OVxHIQ

None of the above

Reconnected to Message Server!

As the author both of this book and of many end-user–interface programs, I find myself shaking my head in disbelief at the horrible user interfaces I encounter from time to time in other people's programs. It is extremely tempting, after a few consecutive successes, to believe that I would never subject my users to similar indignities. Such arrogance, however, is never well-founded. The worst parts of user interfaces are usually the parts the designer simply failed to think about properly. Here, several layers of failure in my software have conspired to produce an error that doesn't match any of the known error codes, and hence cannot be properly diagnosed and explained to the user. But even in this worst-case situation, I could have done a better job than this. My user, predictably, was not amused.

Epilogue

Programming, Humility, and the
Eclipse of the Self

The infinitely little have a pride infinitely great.

—Voltaire

If you build user interfaces and your intuition is good, you'll be lucky to get things right about a quarter of the time, and you won't necessarily know which times you were right.

If you build user interfaces and your intuition is bad, you'll get things right even less than a quarter of the time, but you'll probably think you're doing much better than that.

If you follow the principles described in this book and the procedures outlined in the last chapter, your design skills and intuition won't improve one bit. You'll still be wrong a lot more often than you're right. But you may stand a somewhat better chance of being able to tell, after the fact, what you did right and what you need to try again. In order to do that, you will need one thing more than anything else: *humility*.

You will need humility to recognize that the users are always the final and most appropriate judges of your work. Even though you're the expert in interpreting your users' comments and meeting their needs, their judgments of your work are final and without appeal.

You will need humility to recognize the transitory nature of everything you do. To do your job best, you must recognize that all of your code will ultimately be replaced. Frederick P. Brooks, in *The Mythical Man-Month*, made many programmers aware of the value of prototyping for the first time when he said, "Plan to throw one away.'' Realistically, however, you must plan to throw them *all* away in the long run—a humbling prospect.

You will need humility to be able to take advice from the nearly infinite variety of users and experts who may be able to help you improve your interface.

You will need humility in order to be ruthlessly honest about your own work. You must be able to recognize and admit it when your efforts have led you down blind alleys, and your work must be thrown away almost as soon as it is written.

You will need humility in order to force yourself actually to keep lists of everything that is *wrong* with your system, so that you or someone else can improve it in the next release.

Designing user interfaces is extremely challenging and richly rewarding. There is a special thrill that one gets from seeing strangers using and enjoying the program that one worked so hard to write. It is, in the main, an anonymous thrill—great interface designers are rarely stopped on the street to give autographs—but it is a kind of satisfaction that would be familiar to generations of artisans since before the dawn of civilization.

This familiarity is no coincidence. Despite all the novelty and complexity of modern computing, the skills involved in good interface design are not fundamentally new ones, but draw on millennia of human experience in designing and building artifacts. Like all good artisans, interface designers should focus all of their specialized skill, training, and experience on producing an artifact that is so well suited to its intended use that the end user will never once stop to think about the talented individual or group of individuals who worked so hard to create it.

The greatest compliment that can be paid to an interface designer is for users to regard the interface as so natural that it never occurs to them that any special skill was required to invent it. Interface designers thus toil ceaselessly to make their own work invisible and their own significance largely lost in anonymity. The humility required to embrace such anonymity *as a goal* is the single most important aspect of the designer's art.

| THIS PAGE INTENTIONALLY LEFT BLANK |

Nearly anyone who has read more than a few computer manuals has seen a sentence like the above, sitting alone mysteriously on a page. One presumes that there is a reason for such blank pages, but they are never explained.

    If this page were simply left blank, without the uninformative "explanation" above, this would violate a widespread convention that right-hand pages in books should never be left blank. The authors of computer manuals, however, are rarely concerned with such niceties. Most often, they include the sentence above merely to ensure that empty pages aren't omitted, thus confusing the differential formatting of left-hand and right-hand pages. Thousands of people are thus left to puzzle over the reason for an intentional blank page, simply for the convenience of the person putting together the final version of the documentation. Placing a simple geometric design on the center of the otherwise-blank page would be more "reader-friendly," but it is rarely done.

```
┌─────────────────────────────────────────────────────────────┐
│ ⊠ greenbush ░░░░░░░░░░░░░░░░░░░░░░░░░░░░░░░░░░░░░░░░░░░░░░   ⊡ │
├─────────────────────────────────────────────────────────────┤
│ greenbush nsb 20 % man 3 intro                              │
│                                                             │
│ INTRO(3)            C LIBRARY FUNCTIONS           INTRO(3)   │
│                                                             │
│ NAME                                                        │
│      Intro - introduction to user-level library functions   │
│                                                             │
│ DESCRIPTION                                                 │
│      Section 3 describes user-level library  routines.  In this │
│      release, most user-library routines are listed in alphabeti- │
│      cal order regardless of their  subsection  headings.   (This │
│      eliminates  having  to search through several subsections of │
│      the manual.)  However, due to their special-purpose  nature, │
│      the  routines  from  the  following libraries are broken out │
│      into the indicated subsections:                       │
│                                                             │
│      +  The Lightweight Processes Library, in subsection 3L. │
│                                                             │
│      +  The RPC Services Library, in subsection 3R.         │
│                                                             │
│      +  The System V Compatibility  Library,  in  subsection  3V. │
│         This library contains System V versions of functions that │
│         are not yet merged into the standard Sun  libraries.   To │
│ ▓ --More--                                                  │
└─────────────────────────────────────────────────────────────┘
```

Looking for good books about user interface design can be frustrating. This is the UNIX system documentation's idea of "user-level" information.

Further Reading

And furthermore, my son, be admonished: of making many
books is no end; and much study is a weariness of the flesh.

—Ecclesiastes 12.14

For the serious interface designer, the topic of "further reading" is
like a good news–bad news joke. The good news is that there is a
core of a few good books that will bring you pretty much up to the
state of the art. The bad news, of course, is precisely that there are so
few books that you need to read to get caught up to the state of the art.
Here I seek to provide not an exhaustive or definitive list, but rather a
selection of the handful of books that I consider most valuable for the
aspiring designer.

On the subject of how to design good user interfaces, the two
best books are probably Paul Heckel's *The Elements of Friendly
Software Design* and Donald A. Norman's *The Psychology of
Everyday Things*. Between these two books, you get both an artistic
and a psychological perspective on the design process.

On the more general subject of software engineering, the best
book remains, after nearly two decades, Frederick P. Brooks's *The
Mythical Man-Month*. Although not a textbook, it is considerably
more useful than any textbook of which I am aware.

On the subject of psychological theory of user interfaces, the
literature is not terribly useful, as noted in the text. However, the
interested reader should start with Stuart K. Card, Thomas P. Moran,
and Allen Newell's seminal text, *The Psychology of Human-
Computer Interaction*.

With regard to a better understanding of the statistics that are
often bandied about in user-interface arguments, the reader should be

armed with the skeptical defenses and insights provided so painlessly by Schuyler W. Huck and Howard M. Sandler in their book on the use and misuse of statistical data, *Rival Hypotheses: Alternative Interpretations of Data-Based Conclusions*.

Finally, for realistic insights into the way things are actually done in the computer industry, David E. Lundstrom's memoirs, *A Few Good Men from Univac*, and Tracy Kidder's well-known book, *The Soul of a New Machine*, should not be missed.

References

Borenstein, Nathaniel S. 1985a. "The Design and Evaluation of Online Help Systems." Ph.D. dissertation, Carnegie Mellon University.

————. 1985b. "The Evaluation of Text Editors: A Critical Review of the Roberts and Moran Methodology Based on New Experiments." *Proceedings of CHI.*

————. 1990. *Multimedia Applications Development with the Andrew Toolkit.* Englewood Cliffs, N. J.: Prentice Hall.

————, Craig Everhart, Jonathan Rosenberg, and Adam Stoller. 1988. "A Multi-media Message System for Andrew." *Proceedings of the USENIX Technical Conference.*

————, and James Gosling. October 1988. "UNIX Emacs: A Retrospective." *Proceedings of the SIGGRAPH User Interface Symposium.*

————, and Chris Thyberg. April 1991. "Power, Ease of Use, and Cooperative Work in a Practical Multimedia Message System." *International Journal of Man–Machine Studies.*

Boyarski, Dan, Chris Haas, and Nathaniel S. Borenstein. 1988. "Carnegie Mellon's Andrew: The Evolving User Interface of the Messages Program." Interactive poster session at SIGCHI, Washington, D.C.

Brand, Stewart. 1987. *The Media Lab.* New York: Viking.

Brooks, Frederick P., Jr. 1975. *The Mythical Man-Month.* Reading, Mass.: Addison-Wesley.

————. April 1987. "No Silver Bullet: Essence and Accidents of Software Engineering." *IEEE Computer Magazine*, pp. 10–19.

Card, Stuart K., Thomas P. Moran, and Allen Newell.   1983. *The Psychology of Human-Computer Interaction.* Hillsdale, N.J.: Lawrence Erlbaum Associates.

Carroll, John M., and Caroline Carrithers.   August 1984.   "Training Wheels in a User Interface." *Communications of the ACM*, vol. 27,   no. 8.

Dijkstra, Edsger.   March 1968.   "GOTO Statement Considered Harmful." *Communications of the ACM*, vol. 11,   no. 3.

————, et al.   December 1989.   "A Debate on Teaching Computing Science." *Communications of the ACM*, vol. 32,   no. 12.

————, et al.   May–December 1987.   " ' "GOTO Considered Harmful" Considered Harmful' Considered Further." *Communications of the ACM*, vol. 30,   no. 5–12.

Draper, Stephen W.   March 1984. "The Nature of Expertise in UNIX" (Technical Report).   University of California at San Diego: HMI Project.

Finin, Timothy W.   1983. "Providing Help and Advice in Task-Oriented Systems." *IJCAI Proceedings.*

Heckel, Paul.   1991. *The Elements of Friendly Software Design.* Alameda: Sybex.

Howard, John H.   1988. "An Overview of the Andrew File System." *Proceedings of the USENIX Technical Conference.*

Huck, Schuyler W., and Howard M. Sandler.   1979. *Rival Hypotheses: Alternative Interpretations of Data-Based Conclusions.* New York: Harper and Row.

Kidder, Tracy.   1981. *The Soul of a New Machine.* Boston: Little, Brown, and Company.

Knuth, Donald E.   1968 and later. *The Art of Computer Programming,* vols. 1–3. Reading, Mass.: Addison-Wesley.

Lundstrom, David E.   1987. *A Few Good Men from Univac.* Cambridge: MIT Press.

Morris, James H.   February 1988. "'Make or Take' Decisions in Andrew." *Proceedings of the USENIX Technical Conference.*

————, Mahadev Satyanarayanan, Michael H. Conner, John H. Howard, David S. H. Rosenthal, and F. Donelson Smith. March 1986. "Andrew: A Distributed Personal Computing Environment." *Communications of the ACM*, vol. 29, no. 3.

Naur, P., and B. Randell, eds. 1968. *Working Conference on Software Engineering*, West Germany.

Norman, Donald A. 1988. *The Psychology of Everyday Things*. New York: Basic Books.

Orr, Julian E. 1986. "Narratives at Work: Story Telling as Cooperative Diagnostic Activity." *Proceedings of the First Conference on Computer-Supported Cooperative Work*. ACM SIGCHI.

Palay, Andrew J., Wilfred J. Hansen, Michael L. Kazar, Mark Sherman, Maria G. Wadlow, Thomas P. Neuendorffer, Zalman Stern, Miles Bader, and Thom Peters. 1988. "The Andrew Toolkit: An Overview." *Proceedings of the USENIX Technical Conference*.

Parnas, David L. September–October 1985. "Software Aspects of Strategic Defense Systems." *American Scientist*.

Pearson, Glenn, and Mark Weiser. 1986. "Of Moles and Men: The Design of Foot Controls for Workstations." *Human Factors In Computing Systems, CHI Conference Proceedings*.

Roberts, Teresa L. 1979. "Evaluation of Computer Text Editors." Ph.D. dissertation, Stanford University.

————, and Thomas P. Moran. April 1983. "The Evaluation of Text Editors: Methodology and Empirical Results." *Communications of the ACM*, vol. 26, no. 4.

Rubin, Frank. March 1987. "'GOTO Considered Harmful' Considered Harmful." *Communications of the ACM*, vol. 30, no. 3.

Stallman, Richard M. June 1981. "EMACS, the Extensible, Customizable Self-Documenting Display Editor." *Proceedings of the ACM SIGPLAN/SIGOA Symposium on Text Manipulation*.

Weinberg, Gerald M. 1971. *The Psychology of Computer Programming*. New York: Van Nostrand Reinhold Co.

Wilensky, Robert. Spring 1984. "Talking to UNIX in English: An Overview of an Online UNIX Consultant." *AI Magazine*, vol. 5, no. 1, pp. 29–39.